BUSINESS ECONOMICS

BUSINESS ECONOMICS

Knowledge & Energy

Dr. Hemmat H. Safwat

Printed in the United States of America.

PCIP: Applied

ISBN: 978-1-63385-444-4
Library of Congress Control Number: 2022900710

Published by
Word Association Publishers
205 Fifth Avenue
Tarentum, Pennsylvania 15084

www.wordassociation.com
1.800.827.7903

Contents

ACKNOWLEDGEMENT

The author wants to acknowledge the inspiring discussions that he had with Professor Dr. Ibrahim Oweiss through the journey that started in 1988 and was culminated in the joint completion of the Reference in 2002. Valuable exchanges continued in the early 2000's.

Unfortunately, when the author embarked on the preparing the manuscript of this book, he lacked the insights of Professor Oweiss.

The author wants to present this book to his late wife Mouchira who had always supported him and sacrificed a lot through out his professional journey. The support and encouragement of his daughters Nehal and Amira has been a continuation of Mouchira's. Also, the encouragement of his brother and sister helped him in sustaining the effort to complete this book

The author wishes to acknowledge useful discussions with Dr. Sherif Danish.

The author acknowledges the extensive use he had from Wikipedia.org. in preparing sections 3 and 4 of this book. The sites were extremely useful. Other sites were also helpful. The author apologizes if he overlooked noting any site, that was not intentional but merely because of problems he encountered due to loss of electronic files for the notes he had compiled.

The author notes the excellent level of the various US Government Agencies that have developed very comprehensive and useful data that can be retrieved on then internet easily and efficiently.

The author notes the excellent editing of the manuscript by Tom Costelo and the final production of the book as composed by Jason Price from Word Association Publishers.

EXECUTIVE SUMMARY

In this book, the author puts forward several concepts and propositions that should revolutionize the thinking of business economics.

A. In all the activities of an enterprise (manufacturing or services), whether (internal or external meaning outsources), the pair of knowledge and energy are present.

B. The enterprise goal is to optimize (meaning minimize) the summation of the inputs of both the knowledge and energy to reach the target output.

C. The origin of knowledge is the thinking from human brains while the origin of energy is nature.

D. Knowledge is a broad term that encompasses wide topics including data, technologies, so is energy which include different types of energy chemical, heat, work, and electricity.

E. The control systems in the electrical generation and the electrical system, with the participation of computers versus the analogous role of the brain in the human body.

F. It is through innovations that new ways are reached which produces better use of knowledge and or energy.

G. For energy as an example in fossil power plants, the principles of thermodynamics have provided foundation for the numerous advances in developing and advancing power generation. The advanced have come about from innovations.

H. In parallel, innovations have driven the applications of knowledge as manifested in the various technologies that emerged and in introduction of new technologies.

I. The close relation between knowledge and energy and the significant similarities in many of their characteristics has led the author to suggest a novel proposition to use the units of energy to represent knowledge **"IT IS ALL ABOUT JOULES"**.

J. A reduction in energy associated with a technology improvement (knowledge contribution) could serve as an indirect quantification of knowledge in Joules.

K. The author recommends investigating the processing of various information services with the aim of quantification of knowledge in joules.

L. The author suggests a scale for Knowledge Innovation Combination "KIC" that can be related to the thermodynamic concept "Exergy". He has begun investigating this analogy building on the Reference.

M. Trade is essentially an exchange of knowledge, even from the days of bartering.

N. Business economics is all about mastering inputs of knowledge and energy combinations to reach output in most efficient way through minimizing the sum of the combinations.

The aim of the book is through deep understanding of key factors affecting business economics, enterprises would be in able better achieve overall targets encompassing. key purposes such as:

- v The customer is the center of attention in businesses. Quality of products and services is top priority.
- v Efficiency is a must as businesses compete.
- v Environment protection, particularly attention to CO_2, is behind global warming.

For an enterprise, it needs to plan and develop its facilities (phase 1) i.e., the set up necessary to offer its products and or services. This is followed by the implantation phase (phase 2) where all requirements for the start of operation are completed. Then comes the operational phase. during the operational phases that is the income producing (phase 3). Throughout all these phases the activities or tasks are accomplished through knowledge and energy. The book starts with going in part I reviewing information related to energy and knowledge. In part II, the qualitative model for the knowledge & energy is introduced with applications of the model to manufacturing and services enterprises. In part III, relevant information from Macroeconomics that shape the environment the enterprise functions under. Additionally, information related to securing the needed investments are covered. In part IV, the novel proposition of the quantitative knowledge and energy model. Proposition to deal with knowledge in Joules (the units commonly applied for Energy) is introduced under **IT IS ALL ABOUT JOULES**.

BRIEF ELABORATION

Section 1 INTRODUCTION covers Background, Organization of Manuscript, Overall Perspective- Additional Introductory Notes

This book builds on a previous 2002 publication by the author and Professor Dr. Ibrahim M. Oweiss, *Economics: New Horizons, Shifting the Paradigm*, **Reference.** The book contained four parts:

Part 1 On the Analogies between "Economics" and "Thermodynamics"

Part 2 Selected Topics, A New Approach to "Production Theory", Capital Engine, Growth Theory, Innovation – Why?

Part 3 Simulations

Part 4 Human Model – What Next

This book has four parts, in which the latter parts build on the information presented in the preceding parts.

PART I RELEVANT INFORMATION (SECTIONS 2, 3 AND 4)

Section 2 THERMODYNAMICS CONCEPTES & POWER SYSTEM covers Thermodynamics Principles, Power Plants (Power Generation) Electrical Power System, Energy Storage, Energy and Transport

The principles of thermodynamics that formed the foundation of the evolution of the fossil power plants that form the backbone of the power generation in the electrical systems of today. The electrical power system starting from power plants (generation), transmission and the distribution to the consumers is discussed. The importance of the controls of these systems are highlighted. Many types of power plants and subsystems in the electrical system are presented. Energy storage of electrical energy. Also, energy in transport systems is presented. Fossil fuels dominate applications in transport; however, electricity use is gaining a lot of momentum amid the increased concerns about the environment. **The information in this section underlines the difference in utility of electrical power "work" – Premium Energy as it has versatile uses, in comparison to "heat".** Also, the information underscores the maturity of the electrical industry as well as the importance of considerations of efficiency, heat rejected from the plant. The design life and the impact to the environment are also addressed.

Section 3 HUMAN BODY covers Body Systems, Brain, Individuals and Families – as the Ultimate Consumers

Analogous to the information of a power plant in section 2, comparable information for the human body is presented. Highlights the human body systems are presented. The sources "food" of energy for human body functioning is covered. After the body systems are described briefly, Nutrition is discussed followed by description of human activities. The efficiency of the use of energy (from food intake), to support the work the body needs to perform its daily activities and life, is noted. The brain as the organ that has control of the human systems is briefly introduced. The brain is behind the control functions of the body, (in comparison to functions of the control system in the power plant in section 2. The energy consumption for the brain is highlighted. The information of this section exposes "where the knowledge originates; "humans". The end-customer is really the target of the enterprises in their activities, i.e., providing the products and services for the end customer, directly or indirectly through intermediates.

While sections 2 and 3 appear to pronounce energy information, they also underline the role and the importance of knowledge.

Section 4 KNOWLEDGE & IMPORTANT TECHNOLOGIES covers Knowledge, Innovation, Example Technologies

The term knowledge encompasses several topics, education and definitions of information and data. Also, it covers the definitions of very important terms e.g., technology and knowhow. Innovations and its relevance to inducing advancements that produce improved pairing of knowledge and energy is discussed. The ponderings refer to the initial ideas on the innovation key role in advancements which was discussed in the **Reference.** To complement the information presented in sections 2 and 3, brief highlights of selected technologies as means of adaptation of knowledge are presented. It is noted that each of the selected technologies is a major field, and the few brief paragraphs cannot by no means encompass the intricacies of the technology, it's history, evolution etc. The selected technologies play major roles in a variety of industries. The briefs reflect status of these technologies.

Example Technologies: *Computer, Controls in Electrical Systems, Communications / Information & Communications Technology, Internet & Cloud, Search Engines, Social Media, Artificial l Intelligence (AI), Robots / Robotics*

The highlights presented in this section are meant to bring attention to the roles of knowledge and energy associated with these technologies.

The information in part I brings relevant information for energy & knowledge that represents the foundation of a model for the combination (pairing) of the knowledge & energy in Part II.

PART II KNOWLEDGE & ENERGY APPLIED TO INDUSTRIES AND SOCIETY (SECTIONS 5 AND 6)

Section 5 KNOWLEDGE & ENERGY covers Knowledge & Energy Interplay, Project Phases, K&E in Electrical System, K&E in an Enterprise to Knowledge and Energy

The cornerstone for this book is the model for "**the interplay of knowledge & energy**". Knowledge & energy represent inputs with the objective of attaining an optimized output.

The conceptual setup of the Knowledge & Energy Interplay model is introduced. This is referred to as the qualitative K&E model. Later in section 9, the quantitative K&E mode is introduced after further insights are highlighted in sections 6-8. The application of the qualitative K&E model is demonstrated in the phases of execution of projects. Having discussed the electrical system in section 2, it was decided to have the first demonstrative application K&E model to the electrical system. The generalization of the application of the K&E model for an enterprise follows. Having the right inputs for the application of the K&E model is underlined.

Section 6 INDUSTRIES & SOCIETY covers K&E Model Application to Operations of Enterprises. Historical Data, Planning and Management, Additional Costs, Government, Society – Customer Attitudes, Recent Achievements & Ongoing Digitization, Knowledge Prominence

The application of the K&E model introduced in section 5 is demonstrated, considered are a few private enterprises in both manufacturing and services. The trend of continual growth of the percentage of services versus the manufacturing in the national economies is noted. Data and information inputs are very important to modeling, one key place for this (if not the most important source) is the repository of the historical data. For execution of processes / tasks, it is necessary to include the time, task durations, and sequence to reach the objectives in a timely manner – that is planning and management. Additional factors in the costs not explicitly in knowledge & energy. These have indirect influence on the outputs. The influence of government on various aspects of the enterprise operations that need to be considered. Further, the influences of the society attitude the enterprise needs to attend to as customer behavior is always subject to changes. Some achievements of "how knowledge & energy combinations" succeeded in reaching good results are presented. The developments in technologies such as those discussed in section 4, digitization has taken great importance. Knowledge has and continues to dominate industrial enterprises. as services and exceed 70 % of the total contributions from the industries to the GDP in many nations.

PART III MACROECONOMICS AND INVESTMENTS COMBINED (SECTIONS 7 AND 8)

This part is devoted to conventional economics and investments. The contents of this part are meant to bring the attention of the reader to information he/she may already know for quick look-up

Section 7 MACROECONOMICS covers Aggregate Supply and Demand, Economics Data, Energy, Labor, Fiscal and Monetary Policy, Personal Income, Prices Index, Competition & Monopoly,

Government, and the Economy, The Future "Effect on the Labor Force", Currency & Exchange, Conflicts, Sanctions, Embargos and Wars, Lessons Learned from COVID 19 Epidemic of 2020, Analysis of Industrial Enterprises, Economies of Developing Nations - Comments

The well-known concept of "supply and demand" is discussed and the influence of knowledge & energy as drivers for push or pull effect(s) on the demands in industries is underlined. The contents then move on to macroeconomics. Statistical data for the US, being the leading economy of the world, is chosen for demonstration purposes Relevant energy and labor data are covered as well. Commentary on the influence of competition and monopoly on the economy, and the environment in which the enterprise operates. Additional comments on the role of the Government in inducing a fair environment, and incentives (whenever necessary) to enterprises to serve the society providing products and services.

A note is included on fears that are expressed that knowledge leads to reduced labor jobs and that is detrimental to the society. Statistics for the US international trade are presented. For demonstrative purposes, historical data for Euro – USD exchange rate is presented. A commentary on the effects of conflicts, sanctions, and wars on the economy of a nation is included. Some comments relating to knowledge and energy enabling coping with the massive slowing of the economies of the world as result of the COVID 19 Epidemic that started in early 2020, are included. A very important subsection 7.15 that presents sample US public data for enterprises in different industry classifications. In absence of historical data for a specific enterprise, public data is used in demonstrative application of the quantitative K&E model later in section 9. A commentary on the K&E model possible application in developing nations is included.

Section 8 INVESTMENTS covers Financial Terms, Financing Types, Financing for Projects Abroad, Financial Analysis, Business Risks, Research and Development (R&D), Development of a 50 MW Solar Plant (Example)

Because investments have profound effect on extracting value from knowledge and energy, this section highlights key aspects in business investments. Definitions for key parameters that are used in financing are noted. Possible ways for financing projects are introduced. For a company that is considering undertaking an investment outside its home country, it needs to deal with other considerations beyond what it typically would be accustomed to.To analyze the projections of the returns for an investment for the project, is an essential part that is key to the evaluation of a project in various stages of a project. Financial analysis is used during the feasibility study of the project, then in detailed financial evaluations during the development of the project. Essentially, the cash flow for the capital injections during the construction phase of the project. The projection of the cash flow expenses and revenues during the operation phase of the project follows. During this phase the repayment of the debt and the related interest payments are projected. Finally, depreciation and taxes are dealt with the aim to see the net returns with time (yearly, quarterly, etc.) and of providing the measures of the profitability. For any investments, there are potential risks, and the principals of the project as well as other stake holders(e.g., financial institutions providing loan coverage) all need to take a good look at the possible risks and take measures to contain or mitigate these risks. Going beyond the mitigation of the risks that are discussed, the principals of the project decide on taking insurance coverage, depending on how reasonable the available insurance policies. There are some insurances that are relatively expensive and could cause appreciable detrimental effect on the project returns and the project principal decides to forego.

For an enterprise, particularly large companies it is essential to spend funds to develop new products. The expenditures in R&D could reach significant levels from the company profit, but R&D has significant effect on the future of the company. A demonstrative example for the development of a 50 MW Solar Plant is presented to show the application of investment planning with emphasis on the K&E model.

PART IV COMBINED OUTPUT FROM KNOWLEDGE AND RELATION TO ECONOMICS (SECTIONS 9 AND 10)

Section 9 IT IS ALL ABOUT JOULES covers Bases for Quantitative K&E Model, Quantitative K&E Model, Contributions of Knowledge & Energy in an Activity, Innovation & Knowledge Application, Interdisciplinary, Other Factors from Cost Perspective, Profit and Taxes

After discussing bases for the Quantitative K&E model the model is introduced. **Using equivalent Joules to represent knowledge contributions** is proposed. **Thus, enabling summing the contributions of knowledge and energy. This is a novel model that is being proposed by the author. The author hopes others will find it useful leading to further advancement of the model soon.** The quantitative K&E model provides a tool to measure the combined result from the inputs of knowledge and energy on the output "a product or a service", hence, a measure of the efficiency. Various factors that could be considered in the levels or grades of knowledge and how these are influenced by innovations. With the relevance of innovation, the author suggested a scale for Knowledge Innovation Combination "KIC" to signify the merit of introducing a change in Knowledge Innovation Combination between two situations 1 and 2. The author suggested analogous term in Thermodynamics "Exergy". The investigation of the analogy goes beyond the current objectives of this book. A commentary is included on how interdisciplinary interactions result in improvements and in some cases lead to innovative solutions. Referring to the classical pricing of costs, additional factors that are not directly falling under the K&E model are addressed. Additional considerations over those noted above primarily profit and taxes are addressed.

Section 10 BUSINESS ECONOMICS KOWLEDGE & ENERGY covers Conclusions, Revisit to **Reference,** Multidiscipline, Overall Perspective for an Enterprise, Recommendations

In this section conclusions are summarized. These are followed by brief discussions of the findings of this book and their possible use in conjunction of information previously introduced in the **Reference**. From the various sections of this book, it becomes clear that information from one field can be helpful in other fields. The possible widening of the application of the quantitative K&E model is underlined. Finally, recommendations for further investigations are included.

Section One: INTRODUCTION

At the outset, it is useful to examine the origins of the three words of the title of this book.

> The origin of "energy" (heat) came about as man lighted fire from wood. Later man looked for ways to help him in lifting / moving weights introducing mechanisms (mechanical energy).
>
> Man's search for performing tasks easier to satisfy basic needs led man to compile what he learned in successfully achieving use(s) meant the birth of "knowledge".
>
> With two persons, one had the knowledge to cultivate wheat and the other knew how to plant cotton. When the two wanted to exchange wheat and cotton, bartering started. What was being exchanged was the knowledge of the two persons. Bartering evolved over the centuries into trading, with the evolution of the broad term "economics".

Over the years population growth developments in knowledge and energy have been instrumental in the growth of economies.

In the last 50 to 60 years, especially over the last two decades, three trends emerged and are now prevalent in the ecosystem. These are:

- The customer is the center of attention in businesses.
  The slogan "customer is king" is now part of most businesses culture.
  Quality of products and services is top priority.
- Efficiency is a must as businesses compete.
- Environment protection, particularly attention to CO_2, is behind global warming.

Elaborations on these trends in relation to the book title "knowledge, energy and economics" are discussed in the sections of this book.

This book aims to establish a basis for the quantification of the combined knowledge & energy contributions to reach a desired output, for example in case of a power plant a targeted efficiency. Through this basis, it is hoped (that) broader applications would be developed in various aspects of economics.

The author had a professional career that spanned more than 50 years, covering technical, commercial, and managerial positions in leading companies. In the text, whenever appropriate, he provides comments based on these experiences.

1.1 Background

In 2002, the author, together with Professor Dr. I. M. Oweiss, published *Economics: New Horizons, Shifting the Paradigm*, **Reference.** The book contained four parts:

Part 1 On the Analogies between "Economics" and "Thermodynamics"

Part 2 Selected Topics
A New Approach to "Production Theory"
Capital Engine
Growth Theory
Innovation – Why?

Part 3 Simulations

Part 4 Human Model – What Next

The authors presented an approach that builds on some similarities that they identified between economics and thermodynamics. That initial thinking behind this approach was how to utilize concepts in the science of thermodynamics to explain some known economic theories and practices as found in literature of economics.

Regrettably, Professor Dr. Oweiss was unable to contribute to this book so the author is presenting a fresh view on the approach / the paradigm introduced in **Reference**.

Over the last twenty years, numerous advances transpired in many fields. The author has attempted to build on new information that has evolved to substantiate the merits of some of the initial findings of the **Reference**. Hence to introduce possible refinements that could complement the material in **Reference.**

Reference: *Economics: New Horizons, Shifting the Paradigm*, H.H.Safwat and I.M.Oweiss, Book ISBN 1-891231-71-5, published by Word Association, 2002.

1.2 Organization of this Manuscript

The book is divided into four parts:

- PART I Relevant Information
- PART II Knowledge & Energy Applied to Industries and Society
- PART III Macroeconomics and Investments
- PART IV Combined Output From Knowledge And Relation To Economics

The author borrowed much recently published information to bring closer to the reader some ideas behind the "paradigm shift" in the **Reference**.

The author does not claim any originality for the borrowed information (in the first part of the book).

In some instances, the reader may opt to skip some details and move to other sections. In other circumstances, he / she may want to revert to further details, or he / she could revert to the references and relevant bibliography subsection.

1.3 Overall Perspective – Additional Introductory Notes

A common thread throughout the book can be seen: "input" with the objective to attain "output". This is accomplished through "transformation". The transformation is a result of a process or a group of processes. Resources are required. Please refer to Figure 1.3.1 for a simplified schematic depicting this concept.

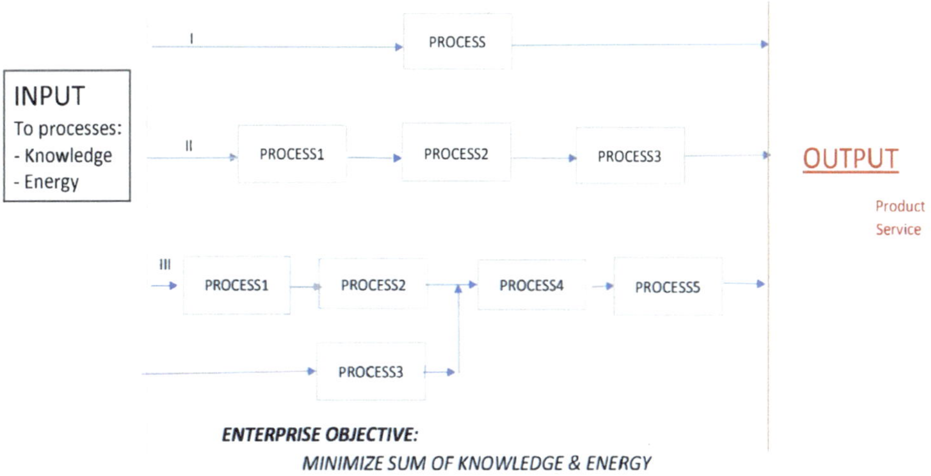

Figure 1.3.1 Schematic for Generic Input – Output for processes of an Enterprise

In Figure 1.3.1 three simple paths for a product / service that is produced in an enterprise. I) is very simple, involving a single process. II) involves several processes that take place in succession (in series). III) involves some processes that are undertaken in parallel as well as series processes.

We always compare the output to the input to get an indication of efficiency. Also, we find that we face inherent losses and/or waste.

The terminology of Thermodynamics – section 2 – contains powerful definitions that can be useful in other fields.

Each of the processes makes use of knowledge and energy input until we reach the output.

The resources are always from knowledge and energy. In a case of a manual process, the knowledge the technician is what he / she is applying. If the person is using an equipment, knowledge is behind the creation of that equipment. Energy was used in the production of the equipment, and electricity used by the equipment consumes energy. This can be generalized to almost all processes in manufacturing and service enterprises.

Further, we recognize always there will be a necessity for the control and management of the processes.

In some cases, we need to resort to storage to match the timings of the use of input to meet the required output.

The applications of thermodynamics (section 2) in the power cycles offer a complete picture of the above perspective. The human body (section 3) practically has the same features.

1.4 References & Bibliography

1) **SAFWAT, H. H**, and **Oweiss, I, M,** *Economics: New Horizons, Shifting the Paradigm*, ISBN 1-891231-71-5, Word Association Publishers, 2002.

PART I
RELEVANT INFORMATION

Section Two:

THERMODYNAMIC CONCEPTS & POWER SYSTEMS

This book aims to deepen the understanding of some close similarities of several aspects of economics to ones in electrical systems, starting from the power generation plants. This section is devoted to introducing to the reader select thermodynamic theoretical principles and concepts in subsection 2.1. These are followed by samples of the the applications of the thermodynamic concepts in subsection 2.2 in typical fossil power plants. In subsection 2.3 how the power plants fit into the electrical system is illustrated. Subsection 2.4 is devoted to energy storage. Subsection 2.5 addresses energy and transport.

The intention is to limit the contents to introductory text that would be simple enough for a non-engineer to comprehend.

2.1 Theoretical Principles

In Thermodynamics, there are three fields: Microscopic, Statistical and Macro. For the purposes of the discussions in this book and specifically in section 2, we shall limit the discussions to topics that fall under Macrothermodynamics.

To start with we introduce thermodynamic properties for a medium (**Fluid**): these are physical parameters, e.g. **Pressure "P", Temperature "T", Volume "V", Internal Energy "I", Enthalpy "H"** and **Entropy "S"**. The latter three properties may not be easy to comprehend at the initial exposure.

To note the units for Energy / Heat / Work being Calories "Cal" or kCal, - British Thermal Unit "BTU" / Joule" J" or kJ and for Power Watt (W) or kW or MW.

A thermodynamic **State** is defined by two independent properties. Thermodynamic **Charts** are utilized to display states and processes. A thermodynamic **Process** represents a change from one state (1) to another state (2). Commonly used charts include P-V, T-S, H-S.

Perfect Gas (ideal) is the simplest medium (fluid) encountered the following relation applies:

$$PV = RT \qquad \text{Eq. (2.1)}$$

Where **R** is **Gas Constant**.

Many mediums including the well-known medium water (**Liquid**) / steam (**Vapor**) are used. For media that has been extensively used such as water/steam– the natural change of **Phase** from liquid to steam provides a useful characteristic as when the two phases coexist. The pressure and the temperature do not define a state (P & T are dependent) and there is a need for an additional property, commonly used "X" **Quality** representing the **Mass** fraction of the vapor to the **Mixture** of the vapor and the liquid.

Two key terms that are used in thermodynamics are **Heat** "Q" and **Work** "W".

A convention in thermodynamics is to use small letters when addressing per unit mass, e.g. "v" which designates volume per unit mass, and "h" is enthalpy per unit **Mass**.

Work is a more beneficial (useful) as through work many actions can be performed while heat does not lend the same versatility. This becomes clear when one sees the extensive steps that are taken for instance in fossil power plants to produce electrical power from the heat released from the combustion of fuel. Electrical energy can be used for numerous purposes.

Two laws govern thermodynamics:

1st Law of Thermodynamics:

In any process, Mass and Energy are conserved. CONSERVATION OF MASS & ENERGY

2nd Law of Thermodynamics:

To convert Heat to Work, only a part of the heat can be converted into work. THIS RELATES TO THE HIGHER-GRADE DESIRABLE FORM OF WORK THAT CAN BE UTILIZED COMPARED TO HEAT.

THIS MEANS THAT IN A CYCLE, NOT ALL HEAT ADDED CAN BE CONVERTED TO WORK (USEFUL). INHERENTLY A PORTION OF THE HEAT ADDED WILL BE REJECTED (NOT USEFUL).

Calorie – Amount of heat needed to increase the temperature of 1 gram of water by 1 Celsius.

Joule - It is equal to the energy transferred to (or work done on) an object when a force of one newton acts on that object in the direction of the force›s motion through a distance of one meter (1 newton meter or N·m). It is also the energy dissipated as heat when an electric current of one ampere passes through a resistance of one ohm for one second.

Conversion Calorie = 4.184 J

Watt - A **watt** is a **unit** of power. A **watt** (W) is a joule (J) of energy used or produced per second.

Thermal **Efficiency** η_{th} is defined as the ratio of the useful "output" W_{net} (work done – internally consumed work) to the "input" heat added Q_A

$$\eta_{th} = W_{net} / Q_A \qquad \text{Eq. (2.2)}$$

Please note that in many locations in this book the word **performance** appears: the two words efficiency and performance are used to convey the same meaning.

2.1.1 Open System

A Thermodynamic **System** is defined as a space subscribed by **Boundaries,** the space inside the boundaries is the interior and outside the boundaries is the **Surroundings**.

The focus is to study the changes in the energy and mass within the interior and follow the heat and work. An open system is characterized by mass flow(s) across the boundaries.

A widely used expression is **Open Steady System** – in this system there is no changes with **Time**.

2.1.2 Closed System

Unlike the open system, a closed system is characterized with zero mass flows across its boundaries. All the exchanges with the surrounding are in forms of heat and work.

2.1.3 Processes & Cycle

A **Process** is defined as a change between two thermodynamic **States**. A Cycle is defined as a sequence of processes that start and end at the same **State**.

Examples of processes are shown in the following Figures 2.1.3.1 & 2.1.3.2, Isothermal – constant temperature and isentropic constant entropy are shown for illustration purposes, using P-V and T-S charts.

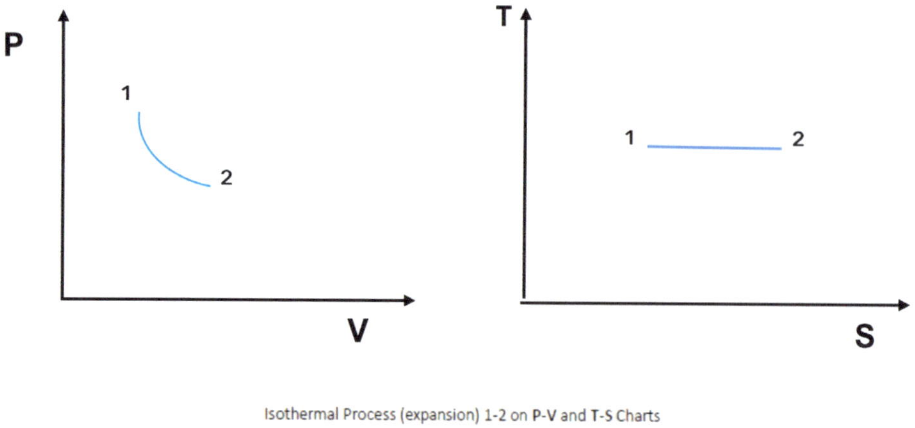

Figure 2.1.3.1 Example of a constant temperature (Isothermal) process between two states 1 to 2.

In the case shown, the process is an expansion moving from high pressure and small volume to a larger volume and lower pressure on the P-V diagram. A compression process would be in the opposite direction. On the T-S diagram the constant temperature is clear.

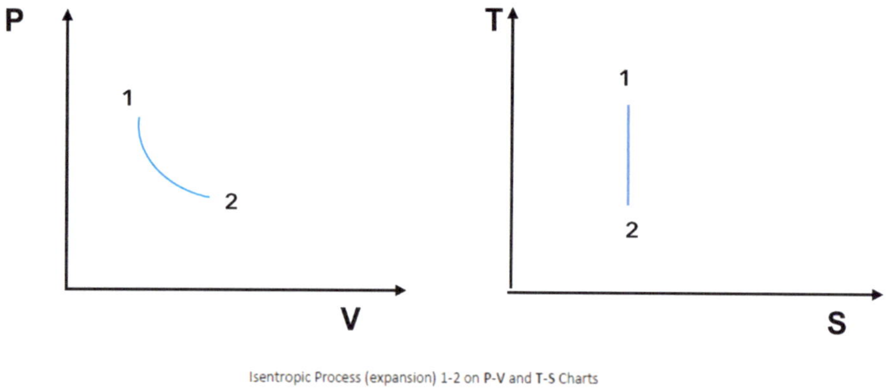

Figure 2.1.3.2 Example of a constant entropy (isentropic) process between two states 1 to 2.

In the case shown the process is an expansion moving from high pressure and small volume to a larger volume and lower pressure on the P-V diagram. A compression process would be in the opposite direction. On the T-S diagram the constant entropy is clear.

2.1.3.1 Carnot Cycle:

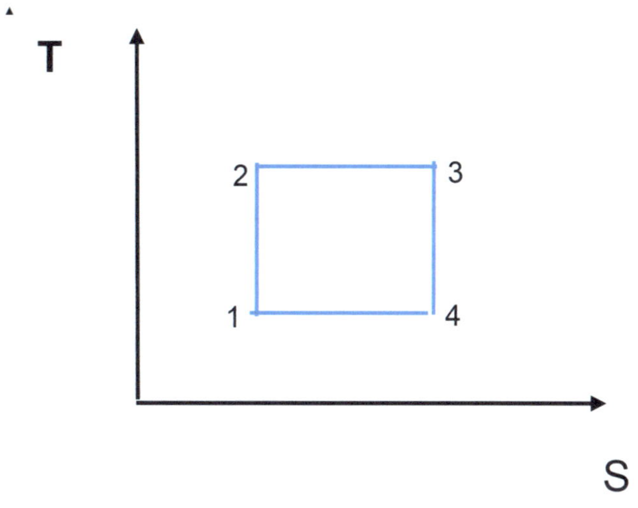

Figure 2.1.3.3 Ideal reversible Carnot Cycle

The Carnot Cycle is an ideal cycle; it is not realizable; however, it is crucial in understanding the second law of thermodynamics. Following the sequence of processes, process 1-2 is an isentropic compression. It is followed by a constant temperature heat addition (Q_A) between 2-3. The expansion 3-4 occurs at a constant entropy. The process 4-1 is constant temperature rejection of heat (Q_R).

$$Q_A = (S_3 - S_2)\, T_H \qquad \text{Eq. 2.3}$$

and

$$Q_R = (S_1 - S_4)\, T_L \qquad \text{Eq. 2.4}$$

The subscripts H and L represents the higher and lower temperatures.

The net work done is equal to the difference between Q_A & Q_R, thus the cycle thermal efficiency per Eq. 2.2 and substituting from Eq. 2.3 and 2.4 for Q_A & Q_R, one gets:

$$\eta_{th} = W_{net} / Q_A = (T_H - T_L) / T_H \qquad \text{Eq. 2.5}$$

Examining Eq. 2.5, shows that in order to attain higher efficiencies we should try to add heat at constant high temperature, the higher T_H the higher efficiency we shall get. Also, we need to reject heat at a constant temperature, the lower that temperature T_L, the higher the efficiency would be. Though the Carnot cycle is a theoretical cycle, it gives us the directions we need to try towards getting the best possible thermal efficiency. Practical challenges are faced in trying to approach the Carnot Cycle:

i) Limits for materials that can withstand high temperatures.
ii) The low temperature is determined by the environment where the cycle is to operate.
iii) Because of irreversibility, it is not possible to realize isentropic processes but what the engineers try to do is approach it as close as possible – overcoming/reducing the sources of irreversibility, such as friction.

The foregoing discussion is fundamental to understanding key ideas behind the theme of this book.

Further one can see by lowering the value of T_L, the efficiency gets larger. In real life applications, the lowest temperature that can be reached is the temperature of the surroundings. This explains why we get more output from the car's engine in winter than in summer time. There are other surrounding factors such as elevation that effect the performance of the thermal engine.

Power plants and engine's development has followed several thermodynamic cycles. Examples of well-known cycles that serve as bases for power plants operation "Byron Cycle" – gas turbines and "Rankine Cycle" for steam and nuclear power plants.

2.1.3.2 Byron Cycle – Gas Turbine

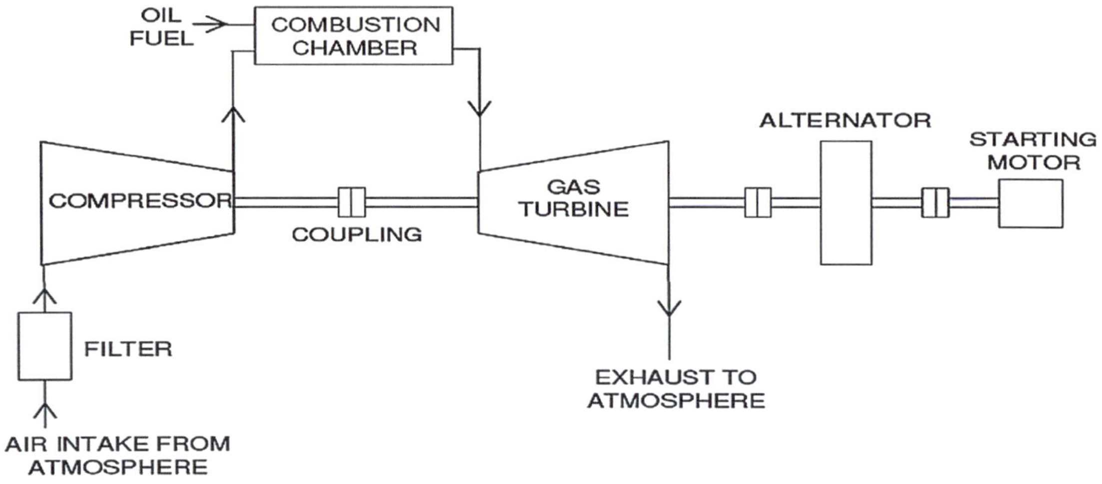

Figure 2.1.3.4 Flow Diagram of Gas Turbine Cycle (Byron Cycle)

The processes that take place for the Byron Cycle of the flow diagram in Figure 2.1.3.4 are illustrated on the T-S and the P-V charts in Figure 2.1.3.5 below.

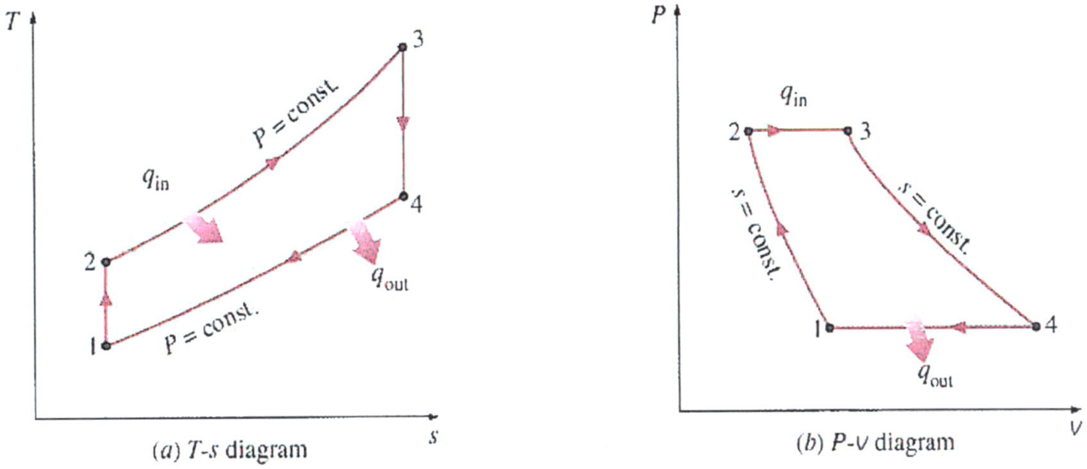

Figure 2.1.3.5 Byron Cycle shown on T-S and P-V Charts

Referring to Figure 2.3.1.5 process, 1-2 takes place in the compressor, process 2-3 (heat addition) takes place at constant high pressure in the combustor; the expansion 3-4 occurs in the two turbine parts, and the constant low-pressure heat rejection occurs through the stack. Without going into further details, it suffices to say that higher efficiency is realized by increasing the pressure ratio.

$$(P_{2-3})\,/\,(P_{4-1}).$$

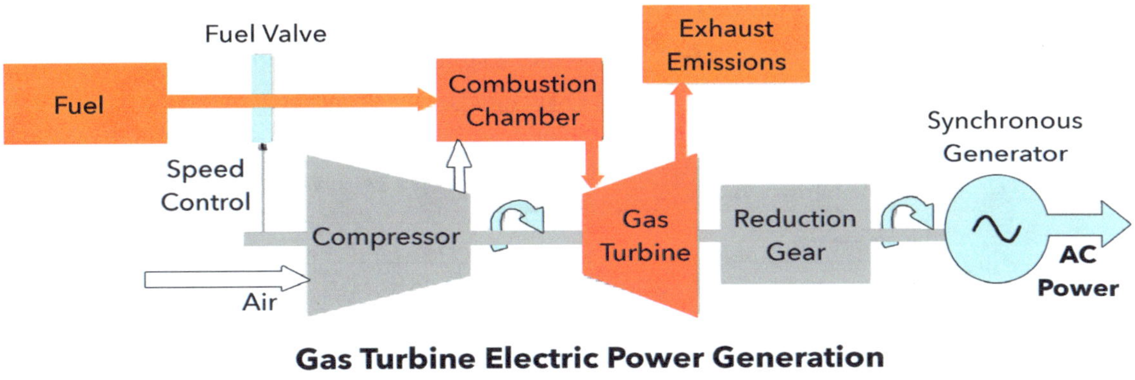

Figure 2.1.3.6 Gas Turbine flow diagram

Gas Turbine

Main Cycle follows Figure 2.1.3.4 and Figure 2.1.3.6

BOP (Balance of Plant):

Fuel Supply and Preparation system (receiving, metering, cleaning, adjusting pressure & temperature),

Compressed Air System, Lubricating Oil System, Water Treatment, Waste Treatment, Closed Cooling System, and Exhaust Gas System.

For Engines, Otto, Diesel and Dual cycles are used – and can be seen in automotive, transport engines.

2.1.4 Heating Value & Fossil Fuels

For fossil fuels, the heating value of the fuel is crucial since in principle the chemical energy released from the combustion of the fuel and hence heat addition in the thermodynamic cycle (in the Combustor in the case of the Byron Cycle – Gas Turbine), and (in the boiler in the case of the Rankine Cycle). Most fuels are hydrocarbons and includes other constituents.

The lower heating value (also known as net calorific value) of a fuel is defined as the amount of heat released by combusting a specified quantity (initially at 25°C) and returning the temperature of the combustion products to 150°C, which assumes the latent heat of vaporization of water in the reaction products is not recovered.

The higher the heating value (also known gross calorific value or gross energy) of a fuel is defined as the amount of heat released by a specified quantity (initially at 25°C). Once it is burned and the products have returned two a temperature of 25°C., the latent heat of vaporization of water in the combustion products is accounted for.

Many Fossil fuels have been used in Fossil Power Plants, depending on the type of fuel; solid (coal), liquid (Diesel Light Fuel Oil, Heavy Fuel Oil) and Gas (natural gas)

FUEL	HEATING VALUE
Hydrogen (H_2)	120-142 MJ/kg
Methane (CH_4)	50-55 MJ/kg
Petrol/gasoline	44-46 MJ/kg
Diesel fuel	42-46 MJ/kg
Crude oil	42-47 MJ/kg

Natural gas	42-55 MJ/kg
Hard black coal (Australia & Canada)	25 MJ/kg
Lignite/brown coal (Australia, electricity)	10 MJ/kg
Natural uranium, in LWR (normal reactor)	*500 GJ/kg*

In the table above, the Uranium was added to show the energy intensity of the fuel of Nuclear Power plants.

Generally, fuel is extracted (mining) e.g. coal mines, then coal is transported to be stored at power plants. For liquid fuels, oil explorations is followed by extraction from oil wells; then in refineries different fuels are produced, then transported to the power plant sites for some storage. For Natural Gas, it is extracted from gas or oil-gas fields and treated before being compressed to be transported in natural gas pipelines to the power plants. In some cases, the natural gas goes to be liquified and then transported mostly in special LNG (Liquified Natural Gas) form to power plant sites where the LNG is stored and re-gasified for burning in the power plants.

2.2 Power Plants (Power Generation)

Over the years different types of power plants have been deployed, different categorization is by fuel, (Coal, Oil, Gas, Waste, Biomass), type Fossil; (reciprocating engines, steam plants, gas turbines, combined cycles), fuel cells, nuclear, and renewable (solar "PV"-photo-voltaic, "CSP" central solar power, wind.

2.2.1 Types

Central Power Generation and Distributed Power Generation.

2.2.1.1 Central

The Power Plant is providing electrical power to a grid, the grid could be a national grid, or part of a national grid (regional) or in small, isolated grids. Isolated grids are found in remote areas or islands. Central power stations have grown in size as larger plants offer economies of scale. However, in some instances the larger size represents some operational drawbacks. We see plants categorized as Base Load (Very Large Sizes), Intermediate (Medium Size) and Peaking (small size).

2.2.1.2 Distributed

Trying to optimize the power generation versus the losses in transmission and distribution by reducing the distances between the location of the power plant and the consumers led to placing some power plants closer to the load centers, all facing the limitations in finding suitable sites for power plants, in cities and to afford suitable means to meet the environmental requirements (emissions, noise, etc.).

Typically, in a Power Plant, there is the main system for converting input energy to electrical energy (output). The electrical energy is the useful type of energy that is sought. The Power Plant has many other supporting systems that are referred to as auxiliary systems including Balance of Plants (BOP) as will become apparent from the examples of power plants discussed below. Very important is the **Control System** that has many important functions for the operation of the Power Plant. Among these functions is adjustment of the Plant Output to meet the required demand and in turn adjusting the working of the various systems (including the auxiliary systems).

Depending on the type and size of the power plant, the footprint (area required for the power plant) is determined. Many factors go in selecting power plant sites that follow candidate sites comparisons.

To be noted generally, solar plants require large footprint as Solar Irradiation energy is dispersed (low irradiation per m2 - 5 kWh/day). Similarly, Wind farms also require large areas as the wind turbines are to be placed at large distance from each other to avoid wake effects from the other turbines. Hydro power plants are characterized by large civil works for the dam and the piping leading to and from the powerhouse.

To be noted, not all the power plants are based on Thermal energy conversions to electrical power; for instance Solar PV (Photo Voltaic) are based on conversion of Electromagnetic waves directly to electrical power in the solar cells. Similarly, wind and hydro convert mechanical energy from Kinetic Energy of the blowing wind and Hydrostatic level difference that becomes Kinetic Energy in the hydraulic turbines.

2.2.2 Fossil Power & Nuclear Plants

Gas Turbines, Steam, combined Cycle (Gas Turbine + steam) and Nuclear Power Plants.

2.2.3 Renewable

Three types are noted below. Renewable plants suffer from inherent intermittent nature. For hydro it is seasonal because of rain collection. And for solar it is the daytime/nighttime plus the cloudiness. For wind, it is the variation of the wind speed and direction. Other renewable plants include bio, geothermal, wave etc.

2.2.3.1 Hydro

Water flows from the high level in the reservoir behind the dam into the Hydraulic Turbine in the Powerhouse. The Turbine drives the electric generator.

2.2.3.2 Solar

Solar irradiation received at the solar panels are converted by the solar cells in the panels to DC (Di rect Current) electrical energy. Through inverters the DC is converted to AC (Alternating Current) to be exported from the power plant.

2.2.3.2 Wind

Wind energy received by the wind turbine's blades rotate the turbine and the generator. In a wind farm typically multiple wind turbines are deployed, and the wind farm has several systems to ensure coordinated energy production.

2.2.4 Cogeneration

Combined Heat & Power "CHP", Desalination and Power and SWRO (Sea Water Reverse Osmosis)

2.2.4.1 Combined Heat & Power "CHP"

In this plant, typically in winter in some locations (cities), heating is required so the waste heat from the Power plant is utilized to provide heating. CHP has found wide application in Commercial settings, e.g., at airports and hospitals. The utilization of the waste heat renders better use of the energy of the burned fuel.

2.2.4.2 Desalination & Power

In many parts of the world the need for potable water has emerged as absolute necessity and "Combined Power – Desalination" has been deployed to provide water for large communities.

SWRO (Sea Water Reverse Osmosis)

In recent years, advances in membrane technologies including substantial reduction of electrical energy used in the membrane have resulted in wider application of SWRO in different parts of the world.

2.3 Electrical Power System

Getting the electrical output of the Power Plants to the Consumers through the Electrical Power System consists of three parts: Generation, Transmission, and Distribution. The transmission part (from generation to distribution) generally includes High Voltage Lines (Overhead or Underground) as well as Substations.

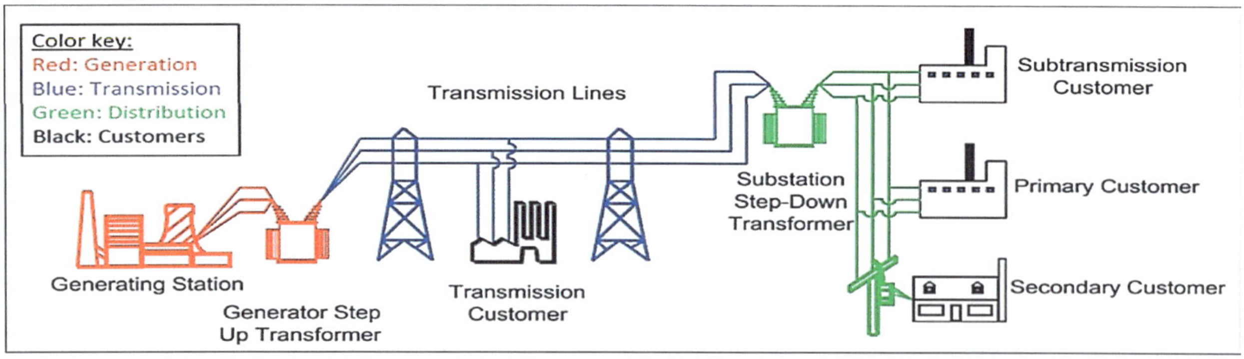

Figure 2.3.1 Typical Electrical System – Generation, Transmission and Distribution

The distribution part starts from the transmission to customers. Consumers are generally characterized as Industrial (manufacturing), Agricultural, Commercial (stores, hotels, hospitals etc.) and Residential. The distribution starts from points in the Transmission and leads to lower voltage lines that feed the Consumers.

All parts of the electrical power system have appropriate functions including controls, security measures, and metering.

With the increased renewable energy (solar PV and wind) – a change in the overall structure of the electrical power systems is taking place, as there is now injection of electricity / multi generations on both the distribution and distribution parts of the networks. There are shifts to digitalization through additions of the large numbers of smart meters to monitor injections and offtake from small producers / users. This increasing trend represents challenges to TSO and DSO (Transmission and Distribution System Operators). TSO & DSO must keep the reliability and quality of standards while facing these challenges.

2.3.1 Generation

Typically, in a fossil fueled power plant – you find the main system – depending on the selected thermodynamic cycle, or a modified form of the cycle plus ancillary systems referred to (balance of Plant) "BOP" including:

Fuel Supply and Preparation system (receiving, metering, cleaning, adjusting pressure & temperature), Compressed Air System, Lubricating Oil System, Water Treatment, Waste Treatment, Closed Cooling System, Exhaust Gas System and Control System.

2.3.2 Transmission

Generally the sites for the power plants are determined by balancing of several factors. Mostly they are some distance from the demand locations – that is the main reason for the Transmission system. A key parameter that faces the planning of the transmission system is the Right of Way of the Transmission Facilities – the Overhead lines and or the underground cables.

As noted earlier, the increased renewable energy is changing the shape of the transmission and the distribution systems as with injection of electricity from smaller power producers – with intermittent characteristics for solar and wind in the transmission and the distribution the typical classical electrical systems in Figure 2.3.1.

2.3.3 Distribution

Figure 2.3.1 Distribution System as interfacing with the Transmission System. The distribution system receives the electrical energy (from the power plants- Generation) via the transmissions system (mostly at a high voltage). In the substations the high voltage energy is reduced to low voltage and delivered to consumers. Distribution systems now see connections of intermittent generation from small or large PV and wind.

2.4 Energy Storage

One can think of energy storage in different forms. Storing the energy in storing fuel is straight forward – e.g., fuel tanks for Diesel (light Fuel Oil) or coal storage at a site of a Power Plant. Storing electrical energy is more complex.

Several systems like Pumped Storage, Compressed Air and Battery storage systems have been in use.

2.4.1 Pumped Storage

In a pump storage system, typically water is pumped during night-time when the demand for the electrical energy is low to the upper reservoir (the Turbo- Electrical Machine) , the hydraulic machine runs as a pump and the electrical machine runs as a motor. During peak hours of the day when the electrical demand in the electrical grid is high, the water flows in the opposite direction driving the hydraulic machine as a turbine and the electrical machine as a generator.

2.4.2 Electric Battery Storage

Battery storage systems have received considerable interest in recent years because of the intermittent nature of operations of wind and PV plants. Also, due to the developments and consumer interest in electric vehicles (see subsection 2.5 below). The developments of batteries are proceeding on different fronts:

- Raw Material; Sourcing / traceability / Life Cycle Assessment (LCA)
- Advanced Material; Li Ion High power / high energy, Li Solid State / Fast Charging

- Manufacturing – Reducing footprint of processes / innovations in machinery and processes / digital twins.
- Smart Functionality - Systems Integration

It should be noted that governments actions have serious effect on the pace of the deployment of battery storage in the electric grids, subsidies, and incentives in dealing with the trading prices of power.

2.5 Energy & Transport

Energy plays a very important role in all types of transport--land, marine and air.

On land transport, motorcycles, cars, trucks, buses, and engines in locomotives of railways have used internal combustion engines (reciprocating engines) for years. These are based on the Otto and Diesel / Dual thermodynamic cycles. They use gasoline or diesel fuel. The engines vary from small sizes (low horsepower) to bigger sizes. In recent years, natural gas has been used to fuel engines. Also, electrical cars and hybrid (fuel + electrical) cars have been gaining market share as the attention to reduction of CO2 intensifies.

Military vehicles and tanks use internal combustion engines.

In marine transport, reciprocating engines power small boats, ferries, medium size, and large ships. For large ships gas turbine engines and jet-propelled drivers have been in use. Mega size vessels for recreational passengers and LNG carriers use bigger drives compared to those in large ships. Older ships used steam cycle plants.

Military navy ships use similar type of drives, and nuclear plants have been used for submarines and naval carriers.

In the air, planes use reciprocating engines and gas turbines (gas turbine (particularly for jet engine on aircraft). The fuel for engines is light hydrocarbon, as the emphasis on weight reduction is a key criterion.

For military air forces, the similar practice is adopted, and in many cases the application in military purposes produce engine refinements that in many cases are adopted in civilian applications.

In recent years, with concerns about fossil fuel emissions, CO2 interest in hybrid cars and in electric vehicles particularly has increased considerably and is expected to grow as battery technology advances further.

2.6 References & Bibliography

1) **Borgnakke, Claus, Sontag, Richard**, (Feb. 2019), *Fundamentals of Thermodynamics*, 10th Edition, Feb. 2019, John Wiley, ISBN: 979-1-119-49496-6
2) **Michael J. Moran, Howard N. Shapiro, Daisie D. Boetner and Margret B. Baily**, *Thermodynamics*, 8th Edition, Wiley, ISBN-13: 978-1118412930, ISBN-10: 1118412931
3) **Stanislaw Sieniutycz & Alexis De Vos,** *Thermodynamics of Energy Conversion and Transport*, Springer Verlag, ISBN10 1461270790, ISBN13 9781461270799
4) **Mohamed Mohamed ElWakil**, Powerplant Technology, Published March 1st, 1984, by McGraw-Hill Science/Engineering/Math, ISBN007019288X (ISBN13: 9780070192881
5) **Mohamed Mohamed ElWakil**. *Nuclear Power Engineering*, New York et al. McFraw Hill Book Company, 1962.
6) **Mehmet Kanoglu, Yunus Cengel and John Cimbala,** *Fundamentals and Applications of Renewable Energy*, McGraw Hill Education, ISBN1 0 1260455300, ISBN13 9781260455304
7) **U.S. Accountability Office** ,*Batteries and Energy Storage : Federal Initiatives Supported Similar Technologies and Goals But Had Key Differences* (GAO-12-842)
8) **Anurag Gaur, A.L.Sharma and Anil Ayra,** *Energy Storage and Conversion Devices: Supercapacitors, Batteries, and Hydroelectric CeAlls*, Taylor & Francis Ltd., ISBN10 0367694255, ISBN13 9780367694258
9) **Ralphp W. Pike,** Continuous Renewable Energy Generation with Lithium Ion Battery Storage on the Micro Grid : A Disruptive Technology, ISBN13 9798571054782
10) **Atle Midtum and Nina Winoszek,** Energy and Transport in Green Transition : Perspectives on Ecomodernity, Taylor & Francis Ltd.
11) **Steveb W. Blume,** *Electric Power System Basics for the Nonelectrical* Professional, John Wiley and Sons Ltd., ISBN10 1119180198, ISBN13 978111918019
12) https://www.researchgate.net/figure/Overview-of-the-traditional-electric-power-system_fig1_321722658

Section Three: HUMAN BODY

It is worth examining how the human body systems work and the role of the brain. The author does not claim to have undertaken in-depth research on this topic as many well-respected scientists and professionals have contributed to the pertinent fields. The literature has plenty of useful articles about the human body and the brain

The goal of this section is to present a brief overview of how the human body systems function. These systems primarily use energy while the brain (center of knowledge) regulates and controls energy use. The perspective of a human engine has several analogous respects to those of heat engines (e.g., power plants) discussed in section 2.

3.1 Body Systems

In the literature, one can find different groupings for the main systems (organs) of the human body. The categorization below is chosen:

a) Circulatory system / Cardiovascular system:

- Circulates blood around the body via the heart, arteries, and veins, delivering oxygen and nutrients to organs and cells and carrying their waste products away.
- Keeps the body's temperature in a safe range.

The circulatory system consists of cardiovascular and lymphatic divisions. The cardiovascular division transports materials within the body and (organs). It includes the heart, blood, and blood vessels.

The lymphatic division (please refer to the below) helps maintain fluid balance and aids in defending the body against infectious diseases. The lymphatic system includes the lymph glands and vessels, spleen, thymus, and tonsils.

b) Digestive system and Excretory system:

- System absorbs nutrients and remove waste via the gastrointestinal tract, including the mouth, esophagus, stomach and intestines.
- Eliminates waste from the body.

These structures and organs enable taking in and digesting food, and absorbing nutrients into the bloodstream for use by the body. The salivary glands, liver, gallbladder, and pancreas facilitate these processes by releasing digestive juices into the digestive system.

c) Endocrine system:
- Influences the function of the body using hormones.

d) Integumentary system / Exocrine system:
- Skin, hair, nails, sweat and other exocrine glands

e) Immune system and lymphatic system:
- Defends the body against pathogens that may harm the body.
- The system comprising of a network of lymphatic vessels that carry a clear fluid called lymph.

Some information has been extracted from different sites; they are included in plates

HOW DOES THE BLOOD CIRCULATORY SYSTEM WORK?

The blood circulatory system (cardiovascular system) delivers nutrients and oxygen to all cells in the body. It consists of the heart and the blood vessels running through the entire body. The arteries carry blood away from the heart; the veins carry it back to the heart. The system of blood vessels resembles a tree: The "trunk" – the main artery (aorta) – branches into large arteries, which lead to smaller and smaller vessels. The smallest arteries end in a network of tiny vessels known as the capillary network.

There isn't only one blood circulatory system in the human body, but two, which are connected: The systemic circulation provides organs, tissues, and cells with blood so that they get oxygen and other vital substances. The pulmonary circulation is where the fresh oxygen we breathe in enters the blood. At the same time, carbon dioxide is released from the blood.

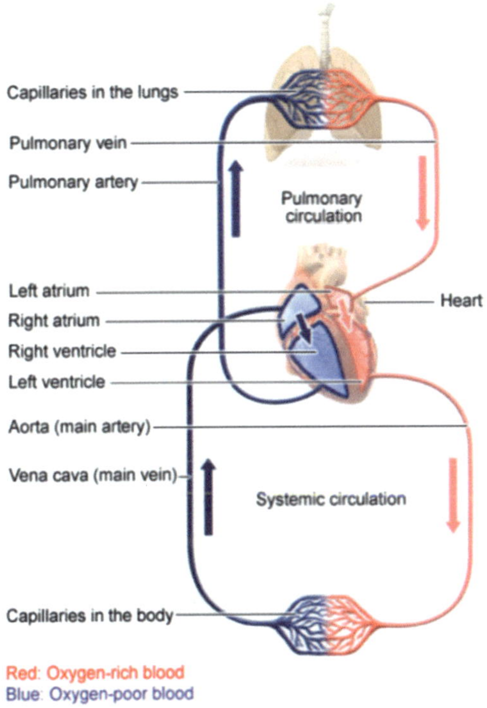

Blood circulation starts when the heart relaxes between two heartbeats: The blood flows from both atria (the upper two chambers of the heart) into the ventricles (the lower two chambers), which then expand. The following phase is called the ejection period, which is when both ventricles pump the blood into the large arteries.

In the systemic circulation, the left ventricle pumps oxygen-rich blood into the main artery (aorta). The blood travels from the main artery to larger and smaller arteries and into the capillary network. There the blood drops off oxygen, nutrients and other important substances and picks up carbon dioxide and waste products. The blood, which is now low in oxygen, is collected in veins and travels to the right atrium and into the right ventricle.

This is where pulmonary circulation begins: The right ventricle pumps low-oxygen blood into the pulmonary artery, which branches off into smaller and smaller arteries and capillaries. The capillaries form a fine network around the pulmonary vesicles (grape-like air sacs at the end of the airways). This is where carbon dioxide is released from the blood into the air inside the pulmonary vesicles, and fresh oxygen enters the bloodstream. When we breathe out, carbon dioxide leaves our body. Oxygen-rich blood travels through the pulmonary veins and the left atrium into the left ventricle. The next heartbeat starts a new cycle of systemic circulation.

https://www.ncbi.nlm.nih.gov/books/NBK279250/

Plate 3.1.1 The working of the Blood Circulatory System

f) Muscular system:

- Enables the body to move using muscles.

The bones, cartilage, ligaments, and muscles of the musculoskeletal system give the body shape, support, and enable voluntary movement. In addition to protecting internal organs, the bones serve as attachment points for your muscles and tendons, produce blood cells, and act as calcium and phosphorus storage banks.

g) Respiratory system:

Takes in oxygen from the air a person breathes and expels other gases such as carbon dioxide, which is a byproduct of cellular metabolism.

h) Skeletal System:

- Bones maintain the structure of the body and its organs.

i) Nervous System:

The brain assembles messages in a way that has meaning for us. It can store that information in human **memory**. The brain controls the thoughts, **memory** and **speech**, **movement** of the arms and legs, and the function of many organs within the body. The central nervous system is composed of the brain and spinal cord.

It is noted that energy (electrical) that the organs use come from chemical reactions in thousands of cells at the parts of the organ from ATPs to ADPs. Adenosine 5'-triphosphate, or ATP, is **the principal molecule for storing and transferring energy in cells**. It is often referred to as the energy currency of the cell and can be compared to storing money in a bank.

THE ROLE OF OXYGEN IN AEROBIC AND ANAEROBIC METABOLISM

The body is dependent upon two processes to create energy at rest and during exercise, aerobic and anaerobic metabolism. Simplified, the aerobic metabolic system uses oxygen, while the anaerobic system does not; however, oxygen plays a vital role in allowing the body to function under both systems.

Both the aerobic and anaerobic systems occur simultaneously; there is no switching from one to the other. Whether the aerobic or the anaerobic system is dominant depends on the amount of muscle oxygen available.

MUSCLE OXYGEN

The term muscle oxygen refers to the balance between the supply and demand of oxygen in the muscle. For example, when demand is exceeded by supply, oxygenation is low, and the muscle requires anaerobic metabolism to provide energy. In the opposite case, when supply exceeds demand, oxygenation is high, and the muscle is able to use aerobic metabolism to power itself.

AEROBIC METABOLISM

Aerobic metabolism fuels most of the energy needed for long duration activity. The aerobic metabolic system functions using the Krebs Cycle, a complex series of chemical reactions that use oxygen to convert nutrients (carbohydrates, fats, and protein) to carbon dioxide and adenosine triphosphate (ATP), an energy-rich compound.

Aerobic metabolism relies on the circulatory system to transport oxygen to the working muscles before it creates ATP. Aerobic metabolism is used primarily during endurance exercises which are typically less intense and can therefore continue for longer periods of time.

ANAEROBIC METABOLISM

Anaerobic metabolism, on the other hand, occurs during short bursts of high intensity exercise, when the oxygen demands of muscle tissue exceed the capacity of the circulation to supply, creating oxygen debt. If the oxygen debt becomes too high, the muscle is no longer able to function. The only way to recover the muscle is to provide it with oxygen-rich blood in order to oxygenate the excess lactic acid. This is usually achieved during a recovery period in a workout.

Many people confuse the concepts of blood lactate and lactic acid. Lactate is produced as a byproduct of the energy system but is completely harmless; in fact, it is actually used as fuel by the heart and muscles. Even at rest, the body produces lactate, though quantities increase with exercise. Lactic acid, however, is a byproduct of lactate buildup and causes acidosis to increase in the system.

THE BOTTOM LINE
Oxygen plays a vital role in both the aerobic and anaerobic metabolic systems. While oxygen is required as a direct energy source in the aerobic system, it is also essential in clearing the blood of excess lactic acid buildup in the anaerobic system.
https://www.business2community.com/health-wellness/the-role-of-oxygen-in-aerobic-and-anaerobic-metabolism-0340309

Plate 3.1.2 Respiratory System

The systems in the human body versus systems in a thermal power plant can be compared. The human body is a lot more complex than a thermal power plant. The human body supports many functions a person undertakes while the thermal power plant aims primarily to produce electrical power from the energy released from the combustion of the fuel.

Note:

The underlined systems (a, b, and g) can be compared to parts in a power plant. As an example, for a) in the case a steam power plant, water/steam & blood are analogous being the medium of the main cycle. However, blood carries a lot of additional functions. For b) the boiler (in a steam plant) performs similar role to that of the digestive system, with the air intake with fans in a boiler and air intake through the respiratory system g) in the human body, The condenser in the power plant has some of the key features of disposing the waste b).

Further, i) the nervous system relates to the brain, in the steam power plant, the control system is responsible for performing many control functions, including starting, stopping, meeting the demand etc. g) the muscular system represents how the human being performs the various functions – while in the steam power plant the ultimate output is electric power from the electric generator.

A steam plant is essentially moving the fluid through different components to achieve specific function, the blood carries out more than one function as it moves in the organs and cells of the body.

The above discussions were limited to the a, b and g systems, the other systems have their own relevance. But for the purposes of this book, the focus is on how the human body gets input, energy (equivalent to heat in a power plant) and produces the useful energy (work) to support all the body systems. In the human body the work is mainly consumed in the internal systems of the body a through j) above, while in the power plant the useful work supports two functions: producing electricity that is evacuated from the plant "Net output", and internal loads to drive the auxiliary systems of the plant as was discussed in section 2.

To complement the above discussions, subsection 3.1.1 below discusses nutrition and subsection 3.1.2 addresses human activities relating to the energy input and outputs in the human body. Subsection 3.2 presents the functioning of the brain controlling the functioning of the human body.

It is worth noting that the human body systems and the brain provide reasonable storage capabilities that the electrical commonly used generation plant (pure plant) does not offer.

The life of the human is not in the person's hand. In a power plant, facilities and equipment deployed are designed for lifetime. Without going into philosophical issues, while the human body has within the systems the capability to support the growth from the birth onwards, the systems in the electrical system have same configuration from the inception onwards. The maintenance and the operation of the human body and the parts of the electrical system have similar considerations.

3.1.1 Nutrition

Food Energy

Food Energy is chemical energy that humans derive from food through the process of cellular respiration. Cellular respiration may either involve the chemical reaction of food molecules with molecular oxygen (aerobic respiration) or the process of reorganizing the food molecules without additional oxygen (anaerobic respiration). Humans need a minimum intake of food energy to sustain their metabolism" and to drive their muscles.

Metabolism is the set of life-sustaining chemical reactions in organisms. The three main purposes of metabolism are: the conversion of food to energy to run cellular processes; the conversion of food/fuel to building blocks for proteins, lipids, nucleic acids, and some carbohydrates; and the elimination of metabolic wastes. These enzyme-catalyzed reactions allow organisms to grow and reproduce, maintain their structures, and respond to their environments. (The word metabolism can also refer to the sum of all chemical reactions that occur in living organisms, including digestion and the transport of substances into and between different cells, in which case the above described set of reactions within the cells is called intermediate metabolism).

Foods are composed primarily of carbohydrates, fats, proteins, water, vitamins, and minerals. **Carbohydrates, fats, proteins,** and water represent virtually all the weight of food, with vitamins and minerals making up only a small percentage of the weight. (Carbohydrates, fats, and proteins comprise ninety percent of the dry weight of foods.) Organisms derive food energy from carbohydrates, fats and proteins as well as from organic acids, polyols, and ethanol present in the diet. Some diet components that provide little or no food energy, such as water, minerals, vitamins, cholesterol and insoluble fiber may still be necessary to health and survival for other reasons. Water, minerals, vitamins, and cholesterol are not broken down (they are used by the body in the form in which they are absorbed) and so cannot be used for energy. Fats and ethanol have the greatest amount of food energy per gram, 37 and 29 kilojoules per gram (8.8 and 6.9 kcal/g), respectively. Proteins and most carbohydrates both have about 17 kJ/g (4 kcal/g). The differing energy density of foods (fat, alcohols, carbohydrates and proteins) lies mainly in their varying proportions of carbon, hydrogen, and oxygen atoms.

Note that the following standardized (simplistic) Table 3.1.1.1 of energy densities is an approximation and the value in kJ/g does not convert exactly to kcal/g using a conversion factor.

Table 3.1.1.1 Energy Density in Food Components

FOOD COMPONENT	**ENERGY DENSITY**	
	kJ/g	**kcal/g**
Fat	37	9
Ethanol (drinking alcohol)	29	7
Proteins	17	4
Carbohydrates	17	4
Organic Acid	13	3
Polyols (sugar alcohols, sweeteners)	10	2.4
Fiber	8	2

All the other nutrients in food are noncaloric and are thus not counted. Compared to fossil fuels, the heating value depends on the chemical composition of the fuel being coal, oil or natural gas, primarily, the hydrocarbons content.

Recommendations for daily intake in the United States are 10,900 and 8,400 kJ (2,600 and 2,000 kcal) for men and women (respectively) for ages of 31 – 35 years. The daily intake depends on the level of activities. Children require higher calorific daily intake (boys higher than girls). A proper mix of foods; bread & cereals, meats & fish, fruits & vegetables, milk and dairy products, and fats & sugar.

In biology, energy homeostasis, or the homeostatic control of energy balance, is a biological process that involves the coordinated homeostatic regulation of food intake (energy inflow) and energy expenditure (energy outflow). The human brain, particularly the hypothalamus, plays a central role in regulating energy homeostasis and generating the sense of hunger by integrating a number of biochemical signals that transmit information about energy balance. Fifty percent of the energy from glucose metabolism is immediately converted to heat.

3.1.2 Human Activities

Human activities are the various actions for living, recreation or necessity done by people. Examples include leisure, entertainment, industry, war, and exercise.

Activities are categorized into four different groups, gestures, actions, interactions, and group activities.

The human body uses the energy released by respiration for a range of purposes: about 20% of the energy is used for brain metabolism, much is used for the basal metabolic requirements of other organs and tissues, and a proportion is used to produce mechanical energy by skeletal muscle to maintain posture and produce motion.

The efficiency of muscles is low: only 18 to 26% of the energy available from food is converted into mechanical energy. This low efficiency is the result of about 40% efficiency for generating ATP(s) "Adenosine triphosphate" – (a molecule that carries energy within cells. It is product of the processes of photophosphorylation that is adding phosphate group to a molecule using energy from light, cellular respiration, and fermentation). ATPs from the respiration of food, losses in converting energy from ATP into mechanical work inside the muscle, and mechanical losses inside the body.

Changes in body temperature – either hotter or cooler – increase the metabolic rate, thus burning more energy. Prolonged exposure to extremely warm or very cold environments increases the basal metabolic rate "BMR" (minimum amount of energy required by body to maintain complete physical mental and absorptive state). People who live in these types of settings often have BMRs 5–20% higher than those in other climates. The type of activities the person undertakes consumes different amounts of energy, e.g. gym activities (training and sports), Outdoor activities, Home & daily life, Home repair, and Occupational activities.

- v It is very important to draw the attention that the body activities use premium form of the energy that is work while the food intake is of the low grade "heat". This is to draw on the notes in section 2. Unfortunately, this fact is not recognized by many and leads to misinformation we can find even by some manufacturers of gadgets and equipment.

3.2 Brain

The brain assembles the messages in a way that has meaning for us. It can store that information in human **memory**. The brain controls the thoughts, **memory** and **speech**, **movement** of the arms and legs, and the function of many organs within the body. The central nervous system is composed of the brain and spinal cord.

The brain is the most complex part of the human body. This three-pound organ is the seat of intelligence, interpreter of the senses, initiator of body movement, and controller of behavior. Lying in its bony shell and washed by protective fluid.

Figure 3.2.1 Shows a very simplified schematic of the brain main parts. The author came up with this diagram as he tried to understand the main structure of the brain and the main functions of these parts.

ORGAN "BRAIN" The human brain is the most complex known structure

NEURONS	**GLIA**
Billions of cells branch-like projections called axons & dendrites, *gather and transmit electrochemical signals*	Billions of cells different types of glilal cells *protect the neurons*

This network of cells enables
Breathing, Loving, Playing, Remembering

Figure 3.2.1 Schematic for the brain cells

Anatomy of the Brain

CEREBRUM

The cerebrum has two halves, or hemispheres, that are further divided into four regions, or lobes. The frontal lobes, located behind the forehead, are involved with speech, thought, learning, emotion, and movement. Behind them are the parietal lobes, which process sensory information such as touch, temperature, and pain. At the rear of the brain are the occipital lobes, dealing with vision. Lastly, there are the temporal lobes, near the temples, which are involved with hearing and memory.

CEREBELLUM

It sits beneath the back of the cerebrum. It plays an important role in coordinating movement, posture, and balance.

DIENCEPHALON

The diencephalon is in the core of the brain. A complex of structures roughly the size of an apricot, its two major sections are the **thalamus** and **hypothalamus**. The thalamus acts as a relay station for incoming nerve impulses from around the body that are then forwarded to the appropriate brain region for processing. The hypothalamus controls hormone secretions from the nearby pituitary gland. These hormones govern growth and instinctual behaviors, such as when a new mother starts to lactate. The hypothalamus is also important for keeping bodily processes like temperature, hunger, and thirst balanced.

The brain stem seated at the organ's base controls reflexes and basic life functions such as heart rate, breathing, and blood pressure. It also regulates when one feels sleepy or awake and connects the cerebrum and cerebellum to the spinal cord.

ANATOMY OF HUMAN MEMORY STORAGE
The parts of the brain which serve as information processors to create memories and store them include the prefrontal cortex, neocortex, basal ganglia, cerebellum, hippocampus, and amygdala. These different parts of the brain have different functions associated with various types of memories.

PREFRONTAL CORTEX

This part of the brain serves its purposes as a store for short-term memory. The prefrontal cortex consists of two functional sides, the left and right. Both sides serve short-term working memory.

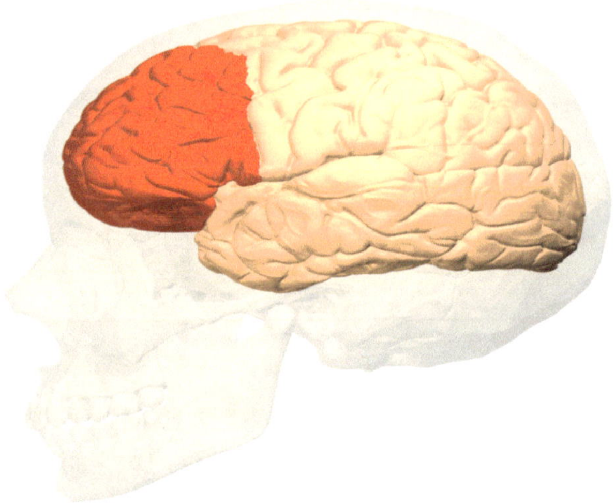

NEOCORTEX

This is part of the brain serves as an information processor. This is the part of the cerebral cortex which is involved in reasoning and learning. It is said that neocortex extracts information from the explicit memories stored in the hippocampus to create reasoning and logic.

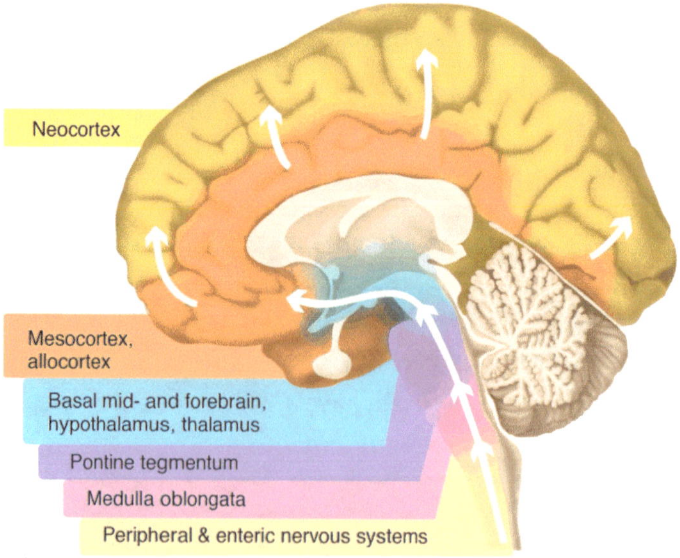

PART 1

BASAL GANGLIA

These are present deep within the brain which keeps the memories that are implicit and provide automatic or involuntary learning from the information provided to the brain in the past.

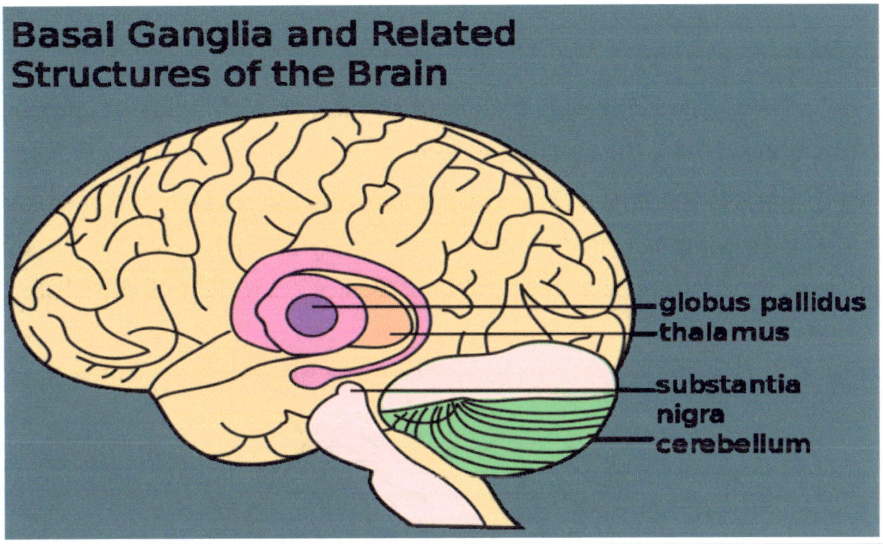

§ AMYGDALA

This is located on the temporal lobe; this is one of the most important parts of the brain which serve for the memory function. Amygdala is known to create sentimental and emotional responses which are then linked to the memories. This helps in the creation of long-term episodic memories.

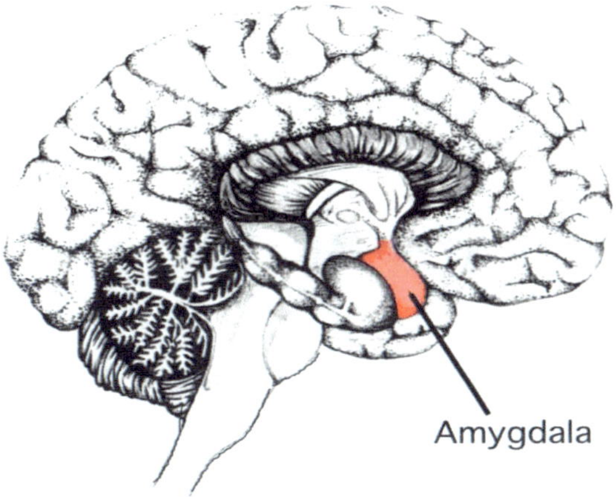

PART 2

HIPPOCAMPUS
The temporal lobe of the brain holds this small structure which holds an enormous list of function. This is crucial for the creation of long-term memories. It acts just like a catalyst for long-term episodic memories. Hippocampus is also essential for short-term memory storage. If the hippocampus gets damaged or surgically removed, like the famous case of Henry Molaison, the brain becomes incapable of forming new memories and store them.
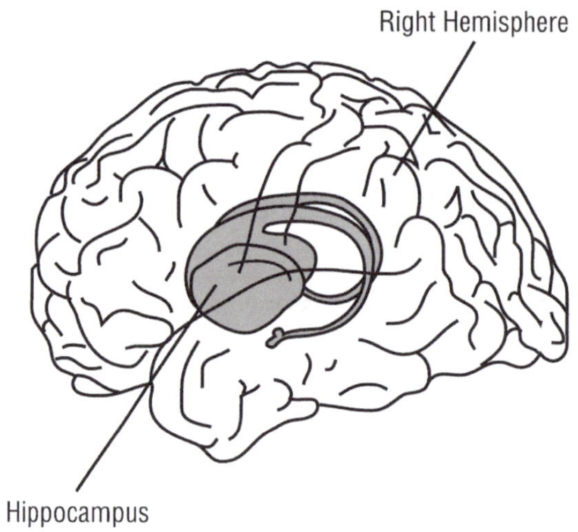
PART 3
https://human-memory.net/memory-storage/

Plate 3.2.1 Anatomy of the Brain

THE FIVE (AND MORE) SENSES

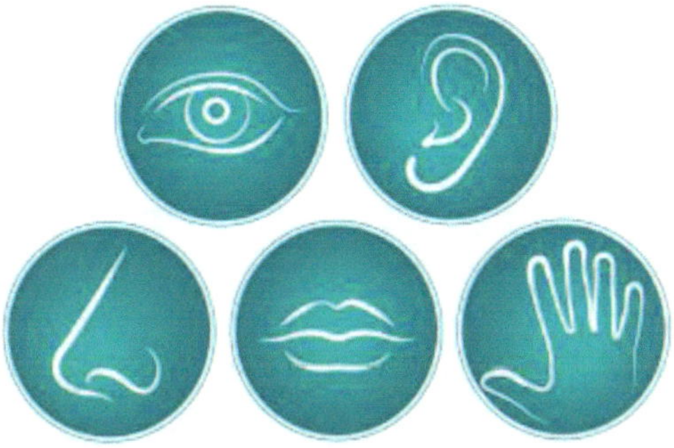

Humans have five basic senses: touch, sight, hearing, smell, and taste. The sensing organs associated with each sense send information to the brain to help us understand and perceive the world around us. People also have other senses in addition to the basic five.

https://www.livescience.com/60752-human-senses.html

Plate 3.2.2 Five Senses

HUMAN MEMORY STORAGE

Memory storage is the process by which the brain can store facts or events so that they can be helpful in the future. It is the process by which life experiences are stored and different skill sets are learned and retained in the brain.

Our brain is continuously involved in the process of memory storage. It receives several pieces of information even within a second, processes them, and stores valuable information in the form of memory.

Memories are stored in the brain at different levels. Those stored at the lower levels are easily forgotten while the ones stored at the higher levels are retained for a longer time.

Our brain acts as a storeroom where memories are stored. Memories related to the things we are more interested in are rapidly stored and remain there for a much longer time.

Different areas of the brain are involved in the process of memory storage. These include;

- Neocortex
- Hippocampus
- Amygdala
- Prefrontal cortex
- Basal ganglia

Based on the time for which information is stored in the brain, memories are of following types;

- Sensory memory is created when sensory information is received and processed by the brain. It is either associated with some previous memory and is stored or is either discarded after its processing.
- Short term memory is the one which is stored in the brain for some seconds to minutes
- Working memory is a type of short-term memory that is considered important by the brain and stored for some time until a task is performed
- Long term memory is stored for months to years. It includes memories about facts and life events as well as memories related to some skills such as writing, typing, etc.

Different models have been proposed to explain the process of human memory storage. The two most famous are the Atkinson-Shiffrin model and the working memory model. A brief detail of both these models has been discussed earlier in the article.

https://human-memory.net/memory-storage/

Plate 3.2.3 Human Memory Storage

It is worth noting the two terms that originate in the brain:

Thought:

an idea or opinion produced by thinking or occurring suddenly in the mind.

Idea:

a plan or a process that occurs in the mind in relation to the completion of a work or duty.

3.2.1 Brain Energy Consumption

Like all the other parts of the human body, the brain needs energy to operate. Answers to three pertinent questions are discussed below.

How much energy does the brain require?

For the average adult in a resting state, the brain consumes about 20 percent of the body's energy. The brain's primary function — processing and transmitting information through electrical signals — is very, very expensive in terms of energy use.

The exact percentages are difficult to ascertain, but we have pretty good estimates of where that energy is going, though it varies by the area of the brain. In the cerebral cortex, about a quarter of the brain's energy goes to maintaining the neurons and glial cells themselves — the processes that all cells go through to remain alive. The remaining 75 percent is used for signaling — sending and processing electrical signals across the brain's circuits.

The bulk of that energy is consumed at the synapses — the tiny gaps between brain cells where signals are sent and received. There, the cells are steadily pumping ions into the gap between cells — exchanging potassium and sodium to create electrical charges. This pumping action is fundamental to the operation of brain circuits, but they are very energy intensive.

APPRAISING THE BRAIN'S ENERGY BUDGET

In the average adult human, the brain represents about 2% of the body weight. Remarkably, despite its relatively small size, the brain accounts for about 20% of the oxygen and, hence, calories consumed by the body. This high rate of metabolism is remarkably constant despite widely varying mental and motoric activity.

The use of modern imaging techniques such as PET and fMRI in the study of the functional organization of the human brain has opened enormously exciting new frontiers in the neuroscience of human behavior. As this work has moved forward, it has become increasingly obvious that understanding the relationship between the signals generated by these imaging devices and the underlying physiology of the brain is critically important to the success and long-range goals of this enterprise. It is heartening that important research work is now emerging on this subject, as exemplified in the two papers in this issue of PNAS. A somewhat unexpected feature of this work more generally is that it is not simply a confirmation of preexisting notions about the relationship between the spiking activity of neurons in the brain, blood flow, and metabolism, but, rather, it is opening up new ways of thinking about the manner in which the considerable resources of the brain are being harnessed in the service of human mental activities. The picture that emerges suggests that neurophysiologists, theoretical neurobiologists, and cognitive neuroscientists, with their imaging devices, all bring important and unique perspectives to an enterprise that is enormously exciting to participants and observers alike.

Plate 3.2.3 Brain Energy Consumption

Do certain parts or areas of the brain require more energy than others?

Between the two major types of tissue in the brain — gray matter and white matter — gray matter requires far more energy than white matter. White matter, made up of bundles of axons, contains large amounts of myelin, the fatty substance that wraps around axons to insulate them and keep electricity from leaking out. Because of this insulation, white matter uses about 20–25 percent as much energy as gray matter, which is made up of dendrites, cell bodies, and the sites of synapses.

Certain functions require more energy than others. The brain areas responsible for auditory processing require more energy than the olfactory system or the areas of the brain responsible for memory. Hearing requires very fast and precise signaling — it wouldn't be advantageous for the sound of danger to be delayed in any way. Relatively slow processes like smell don't have the same intense energy needs.

Does a person use more energy when thinking hard or completing complex tasks?

There are two important factors to keep in mind. The brain requires this expensive electrical power to operate. One's brain never shuts off. Even when sleeping at night, the brain consumes roughly as much energy as it does during the day. While a person rests, the neurons are constantly communicating, updating each other on what is happening. Their constant vigilance is where the bulk of the energy is consumed.

The brain is center of the thinking and the creativity. Ideas originate in the brain of individuals.

3.3 Individuals & Families – the Ultimate Consumers

We should address the ultimate consumer – as individual and or with his / her family. That is the real wheel that turns the economy

Products and services ultimately reach individuals with or without their families. Consumers procure products and services to satisfy their needs or wants. Businesses aim to make these products and goods available to consumers. Some businesses engage in intermediate steps like material and machine parts; for other enterprises the aim is to reach deliver to the ultimate customers.

Some of the services that reach the customer are "shared" meaning these services are not dedicated to the individual, but the service is consumed by groups of people large or small, for instance watching a movie in a theater, or a sports event in a stadium or even traveling in a plane.

The consumer products could be of nondurable products generally having short life (e.g., perishable items such as foods or durable ones such as clothing which have longer life).

While humans use these products and get utility out of them, there is always waste coming out of the usage.

OWNERSHIP

Part of the needs and wants of the individual centers on owning, and this desire of ownership is a behavioral characteristic of humans. Thus, the ultimate consumer (individual & family) is also a consumer that buys long life assets, e.g., houses, buildings, even stock.

3.4 References & Bibliography

1. https://en.wikipedia.org
2. https://www.ncbi.nlm.nih.gov/books/NBK279250/
3. https://www.business2community.com/health-wellness/the-role-of-oxygen-in-aerobic-and-anaerobic-metabolism-0340309
4. https://human-memory.net/memory-storage/
5. https://www.livescience.com/60752-human-senses.html
6. https://www.brainfacts.org/thinking-sensing-and-behaving/learning-and-memory/2018/storing-memories-in-your-synapses-101118

Section Four: KNOWLEDGE & IMPORTANT TECHNOLOGIES

4.1 Knowledge

The word knowledge has been used in numerous instances in the preceding sections of this book. It will appear extensively in the next sections as well. It represents a key word in the theme and title of this book.

Definition

Facts, information, and skills acquired through experience or education leading to the understanding of a subject.

It is clear from section 3, that knowledge originates from the brains of humans. Knowledge encompasses education, information and data, technology, and know-how as briefly presented below.

4.1.1 Education

Education is the process of facilitating learning, or the acquisition of knowledge, skills, values, beliefs, and habits. Educational methods include teaching, training, discussions and directed research.

4.1.2 Information and Data

4.1.2.1 Information

Information is knowledge communicated or received concerning a particular fact or circumstance.

4.1.2.2 Data

Data is facts and statistics gathered for reference or analysis.

4.1.3 Technology

Technology is the sum of techniques, skills, methods, and processes used in the production of goods or services or in the accomplishment of objectives, such as scientific investigation.

4.1.4 Know-how

Know-how is knowledge of how to accomplish an objective effectively with high efficiency.

4.1.4.1 Industrial Know-how

Industrial Know-how is confidentially held, or closely held information in the form of unpatented inventions, formulae, designs, drawings, procedures, and methods, together with accumulated skills and experience in the hands of a licensor firm's professional personnel which could assist a transfer /license of the object product in its manufacture and use and bring about a competitive advantage.

4.1.4.2 Show-how

Show-how is a diluted form of know-how since even a walk-through of a manufacturing plant provides valuable insights to the client›s representatives into how a product is made, assembled, or processed. Show-how is also used to demonstrate technique.

4.1.4.2 Disclosure Agreement

There are two sets of agreements associated with the transfer of know-how agreement: (a) the disclosure and (b) the non-disclosure agreements, which are not separately parts of the principal know-how agreement.

4.1.4.2 Employee Knowledge

Employees have a duty of good faith and fidelity until their employment ceases whereby only the former still applies.

4.2 Innovation

Definition

A new idea or method / the introduction of something new.

Because of the relevance of innovations as they apply to how knowledge applications lead to improvements in various fields, it is important to investigate further different aspects of innovation.

Innovations can be revolutionary leading to a new technology or very large change in a technology. In many instances, patents form the bases for revolutionary inventions. The other form of innovations is evolutionary, meaning they lead to improvement(s).

Innovations lead to attaining better efficiency/performance. Competition among enterprises underlines the importance of innovations to defend the company's market share. The dynamics of the markets drive the competition and hence the emphasis on innovations.

Governments also see the relevance of inducing research in various fields, whether basic research at universities or national research institutes to support the industries.

Referring to the analogy between thermodynamics and knowledge that was touched upon in sections 2 & 3, innovations appear to have levels analogous to what we saw under the second law of thermodynamics--the higher the temperature we can attain better efficiency. The difficulty is that we do not have a good yardstick to adopt for these levels. In thermodynamics, thanks to early researchers, they established the temperature scale.

In the last twenty years many significant advances in different fields have been realized. In this section, we discuss some technologies that have impacted the daily lives of all of us, whether on the business or personal levels. The main driver for these advances is knowledge with the effects that come about as a result of INNOVATIONS. **Innovations are the prime factor that leads to new benefits from knowledge or improved results.**

4.3 Important Technologies

This subsection discusses sample technologies that have played key roles in the progress of many industries. The discussions are limited to simple representations in order not to burden the reader with heavy specialized information. These technologies fall under knowledge. These technologies use energy. This forms the foundation leading to the introduction of the model of knowledge & energy in section 5.

Each of the selected technologies is a field on its own, with its history of evolution and extensive literature; it is not the intention of the author to go deep into details that could be found in the pertinent field literature. The contents discuss simple basics and highlights salient points related to the knowledge and energy pairing.

As the reader goes through the selected technologies in subsections 4.3.1 through 4.3.7, he / she should take note of the energy that each technology uses. The high quality of the energy "electricity" and its reliability 24/7 are underlined. In many instances, battery storage is needed. The words of the combinations of knowledge and energy noted in sections 2, 3 and 4 will become clear as the reader goes to subsection 5.1 and examines the knowledge & energy interplay model.

4.3.1 Computer

Since the 1960's computers have continued to take major and wider use in industries and various uses at home. Figure 4.3.1 depicts the basic structure of a simple computer. Entry of data (input) and (output) data leaving. Please note the concept of input / output (left / right side) – Figure 4.3.1.

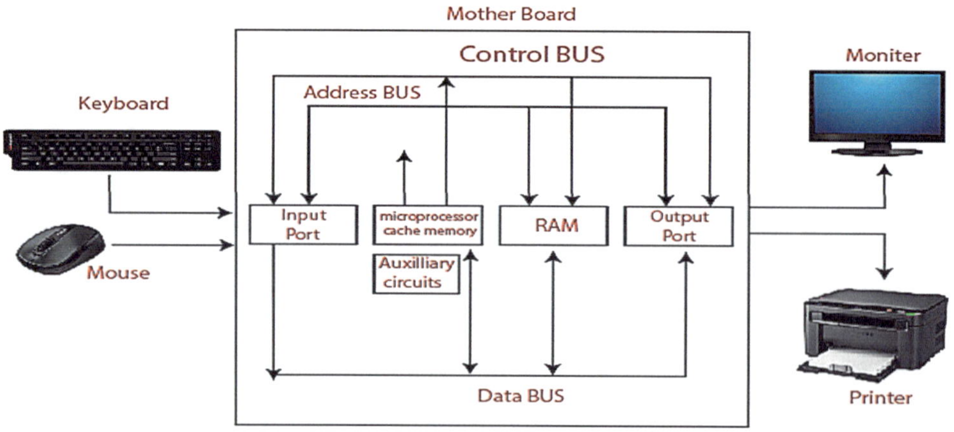

Figure 4.3.1 Architecture of a Basic Computer

Referring to Figure 4.3.1, the Mother Board houses the CPU "Central Processing Unit" which is the main part of the computer. In the figure, the microprocessor is in the center, so is the cache memory and auxiliary circuits, as well as the RAM (ransom-access memory) where data is stored as the processors needs it. In Figure 4.3.1, the Control Bus appears in the upper part of Figure 4.3.1 and the Data Bus is in the lower part. Please note the instructions flow from the main memory to the control unit.

While the bases of the engineering / manufacturing of computer components and their integration are primarily under knowledge, the operation of computers consumes energy. Also, electrical energy storage in batteries is common.

We can see that the applications of computers are working with input data (part of knowledge) and produce as data output another form of knowledge "to meet desired objectives". A key design feature for modern computers is the minimization of the energy consumed.

Laptop computers consume up to 80 percent less electricity than desktop computers and get by on between one-fifth and one-third as much energy. Laptop computers may peak at a maximum draw of only about 60 Watts, whereas common desktops may peak around 175 Watts.

Subsection 4.7, "artificial intelligence" represents an advanced usage of computers to mimic knowledge actions the human brain undertakes in dealing with specific situations.

It is worth noting the similarities between several roles the brain plays in the human body functioning and the analogous functioning of computers.

4.3.1.1 Comparison between the Brain and the Computer

For the comparison, a demonstrative case is considered hereafter. Let's compare the case of an electric security gate that opens automatically when its surveillance camera detects the face of an authorized person like the mailman, and the case of a security guard who opens the electric gate when he sees the mailman. We will use these two cases to describe the similarities and differences between the brain and the computer.

Table 4.3.1 Comparison between the Brain and the Computer

ITEM	SIMILARITIES		DIFFERENCE	
	Brain	**Computer**	**Brain**	**Computer**
1) Input peripheral	The eyes of the guard provide the mailman's face image to the brain as input.	The surveillance camera provides the mailman's face image as input to the computer.	The guard's eyes are directly connected to the guard's brain through optical nerves. There are millions of optical nerves that connect to millions of neurons in the brain.	The surveillance camera is connected to the computer with one cable to the USB port for example. It can be replaced with a newer camera.
2) Processing	The brain uses its neurons and memory to recognize the face of the mailman	The computer's CPU recognizes the face of the mailman from an image stored in its image database on a hard disk for example	The neurons are brain cells that have input "wires" called dendrites that detect the signals coming from the optical nerves. Millions of neurons work together to recognize the face of the mailman by accessing the brain's memory.	Computers typically have one central processing unit (CPU). The CPU receives data from the camera through the USB port and compares it to data stored in the hard disk's database.
3) Output	The brain sends a signal to the muscles in the hand of the guard to press a button to turn on the gate's electric motor to open it	The computer sends an electric signal directly to the gate's electric motor to turn it on and open the gate	The Neurons work together to send chemical and electric signals using their output "cables" called axons to the muscles of the guar's hand.	The CPU sends, through a USB port and a cable for example, an electric signal to the gate's electric motor to turn it on and open it.

4) **Speed**	The mailman is recognized, and the gate are opened, so the task was properly performed.	The mailman is recognized, and the gates are opened, so the task was properly performed,	The guard's eyes send the mailman's face image to the brain, which recognizes the mailman's face from its memory. Then the brain sends signals to the muscles of the guard's hand to press the button. This process takes a few seconds	The surveillance camera recognizes the mailman's face image. It is sent to the CPU, which matches it to the mailman's image in the database. It sends a signal to open the gate at a much higher speed than the human's brain.
5) **Learning**	Brains learn things and store the information in their memory.	Through evolving technologies, Computers learn things and store the information in their memory.	The brain learns much faster than a computer and it stores the information in the form of connections between the neurons. *For example, if you show a few cats to a kid, the kid will quickly recognize any cat when he/she sees one.*	The computer learns by collecting vast amounts information and storing it in its external memory like a hard disk. *For the example of the cats a computer needs to see a very large number of cat pictures in order to learn how to recognize the picture of a cat.*
6) **Memory**	The brain uses the concept of a memory to store information.	The computer uses also the concept of a memory to store information.	The brain's memory is large but is also limited by the brain's biology. *It cannot memorize the phone numbers of an entire phone directory for example. The brain can store 100 terabytes of information.*	Computer memory can be expanded as much as needed. *It can store billions of phone numbers and billions of face images. Computers can store an unlimited amount of information by simply adding more storage devices.*

7) Processing	Brains use the concept of input/process/output to handle different situations	Computers use the concept of input/process/output to handle different situations	The brain relies on billions of neurons that are interconnected by dendrites (for input) and axons (for output) acting in parallel to process input information and output signals to the multitude of body muscles for example.	The computer uses single cable connections from input devices like a camera to the CPU and single cable connection to the output device like a printer for example. The CPU processes the input signals in a serial manner in order to product the output signal.
8) Format of the information			Brains store information in the form of connections between neurons called synapses. Synapses can store the equivalent of multiple terabytes of information.	Computers store information in the binary form of bits that can be either 1 or 0. These bits are stored in tiny transistors or magnetic fields on a hard disk. The numbers 0,1,2,3 are stored in the computer as 00, 01, 10, 11 for example because the computer understands and can process 1s and 0s only. An image is stored in a long series of 1s and 0s. A byte is a sequence of 8 bits. One page of a book contains 3000 letters, each letter is coded in one byte.

9) **Power needed**	Brains take part about 20% of the energy the human body generates from the food intake.	Computers need energy "electricity" to operate	Brains need nutrients and oxygen in the blood in order to function. The equivalent electric power needed by the brain is about 10 Watts	Computers need electric power to operate. A computer needs about 100 Watts to operate. Much higher than the human brain.

Please note that the following subsections deal with technologies that mostly work with / or are deployed via computers.

4.3.2 Controls in Electrical Systems

As was noted in both sections 2 and 3 the controls play key role in the various parts of the electrical system (section 2) and by the brain in the human body (section 3).

In this subsection, we examine an **electrical control** system. The understanding of the working of the electrical control system helps in grasping integration of getting actions completed. The electrical control is a physical interconnection of devices that influences the behavior of other devices or systems. A simple electronic system is made up of an input, a process, and an output. Both input and output variables to the system are signals. Note the similarities between the contents of subsection 4.3.1.

Examples of such systems are found in power plants such as feed water pumps, in air compressors in a gas pump station, or household HVAC system. Sensors gather information about a physical process "cause", then transmit them in an electrical signal towards an output action "effect". The block diagram below illustrates the control system. Two types of control systems are in use: continuous time "CT" signals or discrete time "DT" signals. The two types are an open loop system, or a closed loop system, as depicted in Figure 4.3.2.1 on the following page.

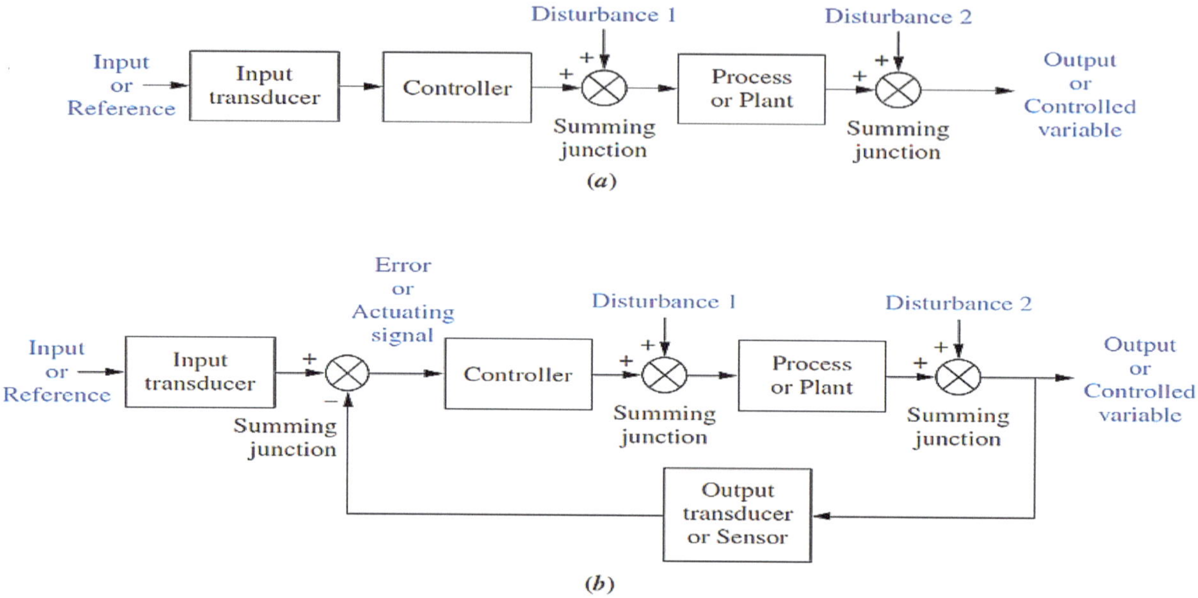

Figure 4.3.2.1 Block Diagram for Control Systems, a) Open loop, b) Closed loop

4.3.2.1 Computer – Controlled Systems

In many modern systems, the controller is a digital computer. The advantage of using a computer is that many loops can be controlled or compensated by the same computer. Additionally, any adjustments of the compensator parameters required to yield a desired response can be made by changes in software rather than hardware.

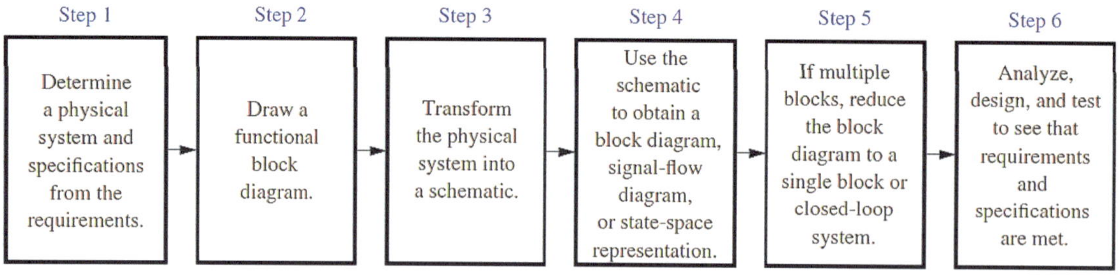

Figure 4.3.2.2 The control system design process

The steps in the design of control system are shown in Figure 4.3.2.2.

Controls of Electrical Systems have grown to be a field in electrical engineering with many books and references in literature. The previous discussions in subsection 4.2 are meant to illustrate the working of the electrical control system. The technology of controls for electrical systems is part of knowledge. In performing the actions mechanical energy is required.

The working of controls is similar to the functioning of the brain in the human body.

4.3.2 Communications / Information & Communications Technology

A communication has three parts: the sender, the message, and the recipient.

The sender 'encodes' the message, "words". A non-verbal communication is transmitted and the recipient decodes it back.

4.3.2.1 Information & Communications Technology (ICT)

ICT is a term for information technology (IT) that underlines the role of unified communications by including the letter C. The integration of telecommunications (telephone lines and wireless signals) with computers smart phones, as well as necessary enterprise software, middleware, storage and audiovisual systems enable users to access, store, transmit, and manipulate information.

ICT covers products that store, retrieve, manipulate, transmit, or receive information electronically in a digital form (e.g., personal computers, digital television, email, etc.).

4.3.3 Internet & Cloud

4.3.3.1 Internet

Internet of things (**IoT**) describes the network of physical objects "things" that are embedded with sensors, software, and other technologies for the purpose of connecting and exchanging data with other devices and systems via the Internet.

4.3.3.2 Cloud

Cloud computing is the on-demand availability of computer system resources, especially data storage (cloud storage) and computing power, without direct active management by the user. The term is generally used to describe data available to many users over the Internet. Large clouds, predominant today, often have functions distributed over multiple locations from central servers.

4.3.4 Search Engine

A **search engine** is a software system that is designed to carry out **web searches** (**Internet searches**), which means to search the world wide web in a systematic way for particular information specified in a textual web search query. The search results are generally presented in a line of results, often referred to as search engine results pages (SERPs).

4.3.5 Social-Media

Social media are interactive computer-mediated technologies that facilitate the creation or sharing of information, ideas, career interests, and other forms of expression via virtual communities and networks. The variety of stand-alone and built-in social media services currently available introduces challenges of definition; however, there are some common features:

4.3.6 Artificial Intelligence (AI)

AI can be defined as development of computer systems able to perform tasks normally requiring human intelligence, such as visual perception, speech recognition, decision-making, and translation between languages, under a wider range of "learning" and "problem solving".

4.3.7 Robots & Robotics

A **robot** is a machine—especially one programmable by a computer— capable of carrying out a complex series of actions automatically. Robots can be guided by an external control device or the control may be embedded within.

4.4 References & Bibliography

1) https://en.wikipedia.org
2) https://www.tutorialandexample.com/block-diagram-of-a-computer/
3) https://www.google.com/search?q=block+diagram+for+a+control+system+open+loop+closed+loop&client=safari&channel=iphone_bm&sxsrf=APq-WBuOAa3yJBBAczLxDQDrhC_1Wb_WgQ%3A1644499229380&source=hp&ei=HREFYpnkFILyaoG9o-gJ&iflsig=AHkkrS4AAAAAYgUfLRW71iIU_0CmK5OOwdh7EgCSMvU2&oq=Block+Diagram+&gs_lcp=Cgdnd3Mtd2l6EAEYADIECCMQJzIECCMQJzIFCAAQkQIyBQgAEJECMgQIABBDMgUIABCRAjIECAAQQzIFCAAQgAQyBQgAEIAEMgUIABCABDoICAAQgAQQsQM6CAguEIAEELEDOg4ILhCABBCxAxDHARCjAjoLCC4QsQMQxwEQowI6CAgAELEDEIMBOggILhCxAxCDAToLCC4QgAQQxwEQowI6BQgAELEDOhEILhCABBCxAxDHARDRAxDUAjoOCC4QgAQQsQMQxwEQ0QM6BQguEIAEUABY6iFgqExoAHAAeACAAY4CiAG6EJIBBjAuMTIuMpgBAKABAQ&sclient=gws-wiz

In part I, we reviewed salient information related to energy -electricity, human body and knowledge elements and select important technologies.

We underlined the importance of attaining best efficiency and minimizing waste.

We noted the life span of the facilities.

We established the similarities between the functioning of the brain in the human body and controls in the power plant.

PART II

KNOWLEDGE & ENERGY AS APPLIED TO INDUSTRIES & SOCIETY

5. KNOWLEDGE & ENERGY

6. INDUSTRIES & SOCIETY

In part II, we present a qualitative model for the interplay of knowledge & energy.

Then applications of this model for manufacturing and services enterprises are presented.

Section Five: KNOWLEDGE & ENERGY

From the previous sections, the close working of knowledge and energy with the control system was underlined in power generation (section 2). Also, in the human body when the brain and the rest of the human body are considered (section 3). Further, we see knowledge and energy combinations under the various technologies discussed in section 4.

A common thread that emerges in all situations discussed in the three sections is the drive towards extracting the largest desired output from any given input combination of knowledge and energy. This will be elaborated upon in section 9 later in this book after covering relevant material in sections 6, 7 and 8, industries, macroeconomics, and investments, respectively.

5.1 Knowledge & Energy (K&E) Interplay

From sections 2 and 3, the two key parameters that affect the working of the main parts of the electrical system (generation, transmission, and distribution) as well as the workings of the human body were identified under knowledge and energy. For instance, from section 2, for a power plant (falling under the generation category), the control system is a key beside the other parameter under energy. A similar conclusion was noted in section 3, for the working of the human body the knowledge centered in the brain, and the energy in other relevant systems in the human body. As noted in both sections 2 and 3, the control system and the brain consume generated energy from the other systems in the power plant or in the human body.

The interplay of knowledge and energy is depicted in the simplified diagram in Figure 5.1.1 below. The contents of this subsection form the qualitative model for the interplay of knowledge and energy "K&E model.

For simplicity, and building on the discussions of section 2, a power plant is chosen to illustrate the twin interplay between knowledge and energy.

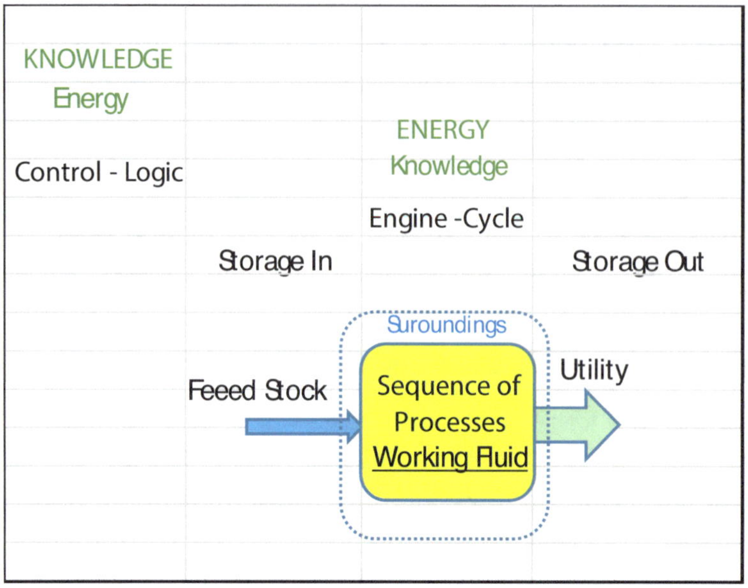

Figure 5.1.1 Diagrammatic Depiction of Inherent Knowledge & Energy Interplay

Referring to Figure 5.1.1 a system– e.g. (power plant) is defined with boundaries – outside the boundaries are the surroundings. Inside the boundaries the power plant systems including the equipment and interconnections – consisting of mechanical and electrical systems, plus the plant facilities and buildings. The main system in the plant is the power cycle as explained in section 2. The importance of the working fluid (in the gas turbine – it is air and the air-combustion products) was underlined in section 2. Other auxiliary systems are essential to the working of the main system. Note the auxiliary systems could consume electrical energy referred to as axillary power. The control system is monitoring and controlling the operation.

In the case of the gas turbine, we are dealing with an "open system" per the definitions in section 2; air and fuel enter the system and exhaust gases (products of the combustion of the fuel) leave the system.

The incoming arrow is the feedstock (fuel). Upon being burned in combination of oxygen from air, the heat is released (in the combustion chamber of the gas turbine). The word utility (output) depicts the "net-work" exported as electrical energy from the gas turbine plant. Note that part of the mechanical power is used to drive the compressor, and another part is used for the auxiliary power. The "net power" is the difference between the power generated in the combustion turbine(s) after taking out the powers consumed in the compressor and the auxiliary power.

In the example of the power plant, the storage is not elaborated on but for generalization, there could be storage of inputs and outputs.

The above discussion underlines a premise that is **the crucial role of the twin inter-relation between KNOWLEDGE and ENERGY the "K&E relationship"**. This is building on the ideas previously put forward by the author and Professor Oweiss (**Reference**). What is always being traded is the value of the knowledge rather than a material content or time spent by a person on one side; and on the other side energy used is the other major parameter in what is being traded.

5.2 Project Phases

It is useful to examine the typical phases that generally are followed in implementing a project that will encompass the infrastructure (yield the fixed asset). The fixed asset will be used to generate the returns for the investments necessary to complete the asset (generally referred to the capital). In Figure 5.2.1 the four phases for a project are shown.

#	PHASE	KNOWLEDGE	ENERGY	OUTSOURCING
1	FEASIBILITY	High	Minimal	Advisors
	PROJECT DEFINITION			
2	DEVELOPMENT	High	Minimal	Advisors
	PROJECT AGREEMENTS SIGNED - FINANCIAL CLOSE			
3	EPC-Cm CONSTRUCTION			
	E -engineering	High	Minimal	Specialization
	P – procurement			
	Supply			
	Equipment – Manufacturing	High	High	Suppliers
	Bulk Material – Manufacturing	Modest	High	Suppliers
	Services			
	C – construction	Modest	High	Subcontractors
	Cm - commissioning	High	Medium	Specialists
	ASSET COMPLETE & READY - COMMECIAL OPERATION			

4	OPERATION & MAINTENANCE	Modest	High	

Figure 5.2.1 Typical phases for undertaking a project "completion of fixed asset"

Under the columns knowledge and energy some qualitative measures on the intensity for the phase or step are noted. These are qualitative and would be different depending on the project nature and unique circumstances.

In Figure 5.2.1, one sees three phases for a project to complete the fixed asset ready to start the production of the project phase 4.

Under phase 1, the feasibility of the project is completed. The feasibility study examines the market, the competition, and projections of the demand with time. Different possible scenarios for implementing the project that would meet the different demand projections are formulated. Estimates for completing the fixed asset are made. These include costs for the land, building, machinery/equipment, and construction. These add to the capital costs for each scenario. Estimates for the running costs to produce certain output(s) "capacities" are made, allowing for costs of input (feedstock and ancillaries), consumables, maintenance etc. By comparing the different scenarios allowing for the cost of money in the investment of the capital costs, the scenario that meets the objectives of enterprise management is selected. Upon getting the internal approvals, additional work is pursued to achieve final project definition to start phase 2 below. Tentative selection of the location of the project is also included.

Phase 2 starts from the main characteristics of the project defined under phase 1 above. It proceeds with the details needed for completing project agreements that will be the basis for the implementation of the project. The details include technical, land lease / purchase, permits, financial and legal. Finalization of the site selection is completed after undertaking necessary site survey(s) and investigations. The technical bases cover all the engineering requirements for the facilities including mechanical, electrical, and civil works. The details of the technical requirements are included in a request for proposal "RFP" for engineering, procurement, construction "EPC". The technical requirements are accompanied by commercial requirements in the RFP. Also, depending on the approach of the principal of the project a bidding may be done to hire an O&M (Operation and Maintenance) contractor. All required permits and consents are secured. Other agreements are negotiated including land lease or purchase and financing package. Depending on the project type some sales agreements are negotiated and signed; for instance under project, finance structure of the critical Power Purchase Agreement "PPA" is negotiated and signed.

Once the foregoing steps are finished, the project agreements are ready and the notice to proceed "NTP" is given to the selected EPC contractor marking the start of phase 3.

Phase 3 is the main phase that translates the results of phases 1 & 2 to the project execution leading at the end of phase 3 a ready "Fixed Asset" that is to attain the objectives for the production from the Fixed Asset. Throughout the four steps of phase 3, strict adherence to quality, health, safety and environmental requirements is a must. Phase 3 has four main steps: engineering, procurement, construction, and commissioning. Based on the EPC contract, the engineering and procurement teams produce all the specifications, drawings, and other engineering deliverables needed by the construction team to complete the construction. The procurement and engineering teams complete the supply contracts for the equipment and bulks, as well as the pertaining third party inspection agreements, in addition to the logistics for marine and inland transport. Application for permits and Environment Impact Assessment EIA are completed, submitted to the cognizant authorities to obtain the applicable authorizations. The procurement and engineering teams also complete and negotiate contracts for the construction packages. In step #3 "construction", the work at the site takes the center of activities under the project site management. The work covers the civil site preparation and then building/facilities construction, mechanical and electrical erection. As step 3 reaches its final stage, pre - commissioning gets under way through inspections of completed equipment / subsystems, etc. The commissioning of the facility under step 4 proceeds to the final testing once the facility achieves the contractual performance; then the fixed asset is ready and the commercial operation under phase 4 begins.

Phase 4 – under this phase the sought production of the fixed asset is fulfilled and the measurement of realization of the commercial objectives over time can be evaluated and judged.

From the discussions in this subsection, throughout all the phases of a project we see the significance of the knowledge and energy – the interplay of K& E discussed in subsection 5.1.

5.3 K&E in the Electrical System

Building on the demonstrative example of Figure 5.1.1, the electrical system is examined further, in this subsection, again from the perspective of the K&E relationship.

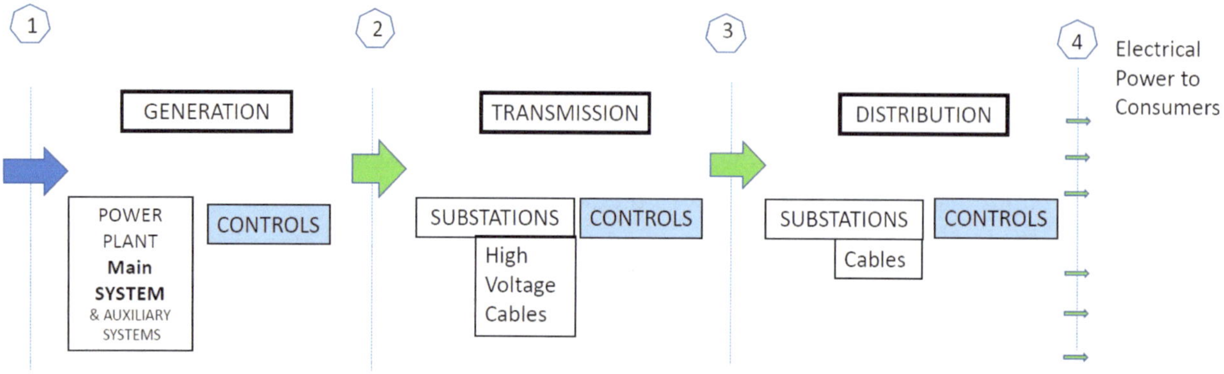

Figure 5.3.1 Depiction of Electrical Systems – Generation, Transmission and Distribution in relation to the K&E model

The simplified schematic in Figure 5.3.1 is representation of the contents of Figure 2.3.1. The purpose is to underline the interplay of K-E in the three parts generation between stations 1-2, and transmission between stations 2-3, and distribution between stations 3-4. In each of the three parts the interplay of K-E is present. An example of the control system can be visualized per the information that was presented in subsection 4.3.2 One can see in the three systems that knowledge is behind the technologies adopted. The design / engineering and completion of the fixed assets, then under operation still knowledge is critical so is the energy supporting the operation.

Let us add to the above the perspective of the efficiency "η", meaning the ratio of the output / input. For the generation part between stations 1-2 the thermal efficiency η is the ratio of the net electrical power to the heat content (added heat) with the fuel feedstock as was discussed in section 2. The efficiency of the transmission part η_T between stations 2-3 is the ratio of the electrical energy passing from station 3 (outgoing) to the incoming through station 2. The higher η_T signifies reduced losses in the components of the transmission part 2. Similarly, the efficiency η_D of the distribution part between stations 3 – 4 refers to the sum of the energy reaching the customers at station 4 compared to the energy entering from the transmission part at station 3. The higher η_D means lower losses between stations 3-4.

5.4 K&E in an Enterprise

The discussions of subsection 5.2 are general and can be applied generically to different types of enterprises with appropriate eliminations / additions. In subsection 5.1, the interplay K-E was addressed. The interplay K&E model was demonstrated for the electrical system (which was elaborated on in section 2), in subsection 5.2. Following the progression of the information presented, hopefully the preceding sections have generated the proper insights for grasping the general concept of the K&E interplay model for other enterprises outside the electrical industry. This is what we are dealing with in this subsection.

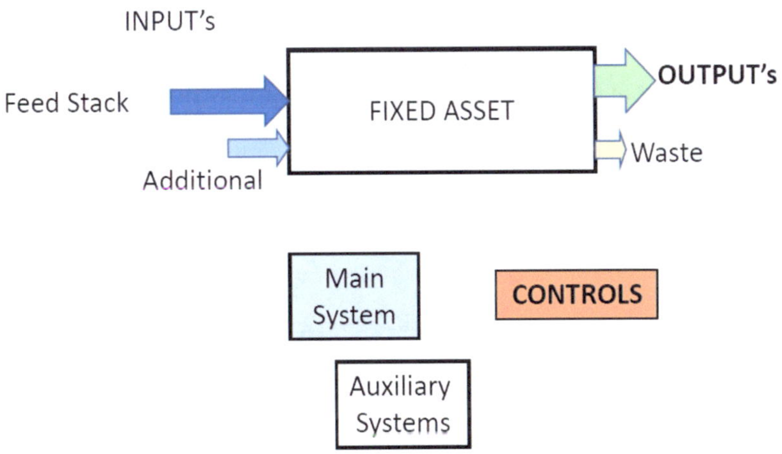

FIXED ASSET: Completed per Project implementation — Figure 5.2.1

OPERATION & MAINTENANCE: Producing the Output's from the Inputs

Figure 5.4.1 Schematic of Typical operation of an enterprise.

The schematic in Figure 5.4.1 is a general one. It is meant to underline the interplay of the K-E in the operation of any enterprise.

The enterprise can be a small / medium manufacturing company, or a large manufacturing company. The fixed asset should be looked at in conjunction with the information in subsection 5.2 ad Figure 5.2.1 (project phases).

The controls box in the Figure signifies knowledge while the main system and the auxiliary systems refer to parts that are dealing with energy. As was pointed out in many parts of this book, in applications we always see knowledge and energy present in the various activities. Repeating the working of the controls needs energy and functions per the designed features that is part of knowledge.

The schematic of Figure 5.4.1 also applies to services. For instance, let us take a doctor that is practicing on his own, the fixed asset is the clinic (completed with the same project phases that are needed to have the clinic ready to receive patients as noted in subsection 5.2). When it comes to the operation, the knowledge of the doctor and the knowledge in the controls of the equipment the doctor uses are part of the knowledge. Electrical energy is being consumed in the equipment the doctor and his assistant(s) use in performing their tasks. Similar rationale applies to the working of a hotel or a financial institution.

5.5 Inputs – Knowledge & Energy

For the K&E interplay model and referring to Figure 1.3.1, for the various processes, inputs for both knowledge / technology and energy are needed – these sometimes are referred to as resources, (please also refer to subsection 6.3 in the next section for planning). For a manufacturing or services enterprise, the energy input to a process is straight forward – mostly electrical energy consumed by the equipment / machine that is used in the execution of the process.

Knowledge input is more complicated. It takes many forms and collectively adds up to the knowledge component entering to that process. Subsection 6.2 "Historical Data" points to possible source for data. Further subsection 4.1 elaborates on the types of knowledge data. As a starting point, we are after the knowledge from brains of the person(s) that is executing the process. In addition, we need to factor in all the knowledge that is indirectly coming from the tools the person uses, whether these come through software or knowhow that was reached through long term practices and/or R&D (internal in the enterprise, or external procured from outside source). The equipment used comes with a knowledge component as well. In the meantime, for getting the software, the know how machines and equipment ready for use input energy was consumed along the noted knowledge input.

The point here is to recognize the fact that knowledge comes in for any process. Also, we can add that the pricing we know now is the only means that we can refer to as a starting point for where to go to get the inputs for knowledge.

Section Six: INDUSTRIES & SOCIETY

It would be useful to review several evolutions since 1960's. This period had many changes which correlate or are closely related to the knowledge / technology and energy. If we go back to 1960's, we see that the US space program was instrumental in developing many new technologies, the aerospace industry made leap steps. Quality assurance started, going beyond quality control. In the 1970's, the nuclear industry emerged and played a major role in realizing mega projects and underlined the quality assurance to ensure the required safety. Manufacturing of heavy and high precision equipment saw broad activities. The construction industry played a major role. By the mid-eighties, movements opposed to nuclear power plants managed to bring the industry to a halt and no new orders were placed by utilities in the US. This was followed outside the US. Public concerns for safety and the disposal of nuclear waste were behind the demise of the nuclear industry. Electrical energy during that era in the US was produced in large coal plants, and in the completed nuclear plants. As a result of the 1973 war in the Middle East, oil and gas started to emerge as sought fuel in many parts of the world. The oil and gas industry saw very large expansion. Petrochemical and LNG industries flourished. Parallel to this large infrastructure projects were undertaken. New suburbs and new cities were built. Construction contractors were involved, with supporting industries providing equipment, and products.

With developments noted above in the energy sphere, many industries made advancements. During this era, many of the leading companies were led by technical / engineers. During this era, we saw many major multinational companies with large portfolios and activities outside their home country. Several industries saw major consolidation as the "production company type" disappeared. Globalization started and it had a profound effect on transforming many of the developing countries from just providing raw material to start working in locally established factories. The employees of these factories have become consumers and consumption worldwide saw big boosts.

In the early 1990's many of the multinationals changed their view on who should be on the helm of businesses, and many of the large companies started operating under non-engineers, realizing the best returns for the shareholders took the front stage for the companies. With this change came a real emphasis on marketing and shift from the production strategy that had grown over the preceding years. Steps to break monopolies and single dominant players were undertaken in some key industries.

New paradigms emerged away from vertically integrated model for companies, that is "out-sourcing; small and medium enterprises" started to prevail. The privatization gradually replaced the public enterprises in many fields throughout the world.

In the 21st century the trends that started in the 1990's continued and as things evolved further, "changes began to be at a very fast pace compared to any time before".

Through the journey noted above the knowledge & energy were at the forefront of all the developments.

With the above brief history, now we start focusing on the application of the K&E interplay model to enterprises. Noting the application of the K&E model of subsection 5.1 was demonstrated in subsection 5.2 for project phases. In subsection 5.3, the K&E model was applied to the electrical system. In subsection 5.4 the model was applied to a generic enterprise.

The intention here is to demonstrate the applicability to various types of enterprises, thus building on subsection 5.4, a select variety of manufacturing and services enterprises are investigated. The focus is on running operations.

In section 7, under macroeconomics, more economic information derived from analyses of economic data are presented. Please refer to section 7 for the make up of the industrial enterprises – services versus manufacturing.

6.1 K&E Model Application to the Operations of Enterprises

In this subsection samples from industries in manufacturing and services, as well as mixed "services & manufacturing" are examined in relation to the K&E interplay model introduced in subsection 5.1. The intention is to focus on the knowledge & energy in the execution processes of the products and services. The K&E influence direct and indirect costs for producing a product or providing a specific service. This is the premise of this book; "that major costs are primarily influenced by the pair K&E".

The additional costs that the enterprise needs to deal with over and above factors on top of the K&E production costs are dealt with in subsection 6.2.

6.1.1 Manufacturing

In Table 6.1.1.0, data from BEA for the 10 years period between 1987 to 1997 are presented for groups of manufacturing industries. At the top row of the table the total value of manufacturing per year is shown in billions of USD. In the two rows below the top row, the percentages of the two categories durable and nondurable products appear, respectively. In the yellowed row, the growth trends over this period for the totals of the durable products and the non-durable products categories can be seen. Then for each year the percentages of various groups falling under the industries for durable / nondurable products have been included. One can see the high growth category in the groups Durable and Nondurable

Table 6.1.1.0 Manufacturing total values in the US for the period 1987-1997.

Line		1987	1989	1991	1993	1995	1997
12	**Manufacturing**	878,808	1,015,795	1,042,999	1,124,330	1,259,569	1,382,889
	Ratio to 1987	*1.00*	*1.16*	*1.19*	*1.28*	*1.43*	*1.57*
	% Durable / Manufacturing	60.6%	59.8%	57.3%	57.3%	57.8%	59.6%
	% Durable / Manufacturing	39.4%	40.2%	42.7%	42.7%	42.2%	40.4%
13	Durable goods	532,481	607,681	597,461	644,598	727,936	823,793
	Ratio to 1987	*1.00*	*1.14*	*1.12*	*1.21*	*1.37*	*1.55*
14	Wood products	19,721	20,606	18,710	22,044	25,641	26,791
15	Nonmetallic mineral products	25,246	27,306	24,310	28,177	33,871	39,766
16	Primary metals	33,015	44,572	37,860	39,703	47,743	48,074
17	Fabricated metal products	69,528	78,167	77,629	83,908	98,789	110,023
18	Machinery	69,562	87,974	83,096	86,079	95,153	102,233
19	Computer and electronic products	98,495	111,819	116,791	124,858	159,672	195,321
20	Electrical equipment, appliances, and components	36,029	41,552	40,785	43,045	44,326	46,856
21	Motor vehicles, bodies and trailers, and parts	71,344	75,458	64,255	89,490	104,158	112,015
22	Other transportation equipment	67,253	67,603	77,468	64,240	52,049	62,539

23	Furniture and related products	17,831	19,635	18,259	21,688	23,174	27,921
24	Miscellaneous manufacturing	24,458	32,990	38,298	41,366	43,358	52,255
25	Nondurable goods	346,327	408,114	445,538	479,731	531,633	559,097
	Ratio to 1987	*1.00*	*1.18*	*1.29*	*1.39*	*1.54*	*1.61*
26	Food and beverage and tobacco products	90,870	100,634	118,455	122,683	136,526	135,203
27	Textile mills and textile product mills	21,126	22,956	23,864	26,967	26,247	27,396
28	Apparel and leather and allied products	22,797	24,811	26,025	27,585	26,771	25,165
29	Paper products	39,280	48,146	46,772	48,781	62,429	55,369
30	Printing and related support activities	26,389	30,412	31,677	33,767	35,655	37,438
31	Petroleum and coal products	21,387	31,452	33,808	38,471	33,878	47,512
32	Chemical products	91,596	110,681	123,947	134,405	158,733	173,408
33	Plastics and rubber products	32,882	39,022	40,988			

It is useful to compare the K&E approach to the classical costing system of the KLEMS. Please refer to the illustration in Figure 6.1.1

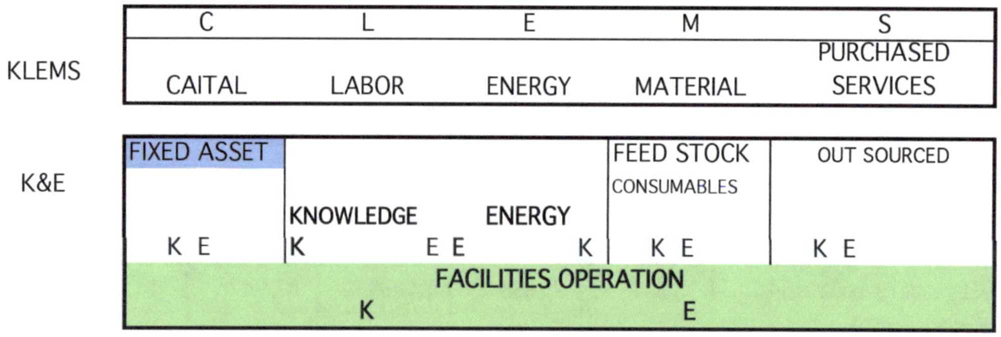

Figure 6.1.1 Comparative KLEMS (top of the Figure) versus the groupings under the K&E model.

For the K&E model the four elements L (labor), E (energy), M (material) and S (purchases of Services) are parts of K&E as will become clear from the examples that follow. The Capital or the costs related to fixed assets also is a result of K&E.

It is necessary to note some considerations that must be factored in applying the K&E model. For instance:

- § The scale of the enterprise.
- § The nature of the enterprise – Multinational, Regional, National, and medium / small enterprise (SME).
- § Multi product / single product.

A cautionary note of the author intention is to show only high-level demonstrations rather than detailed demonstrations at this stage as that goes beyond the objective of this book.

An important question that comes up why the K&E model does not include material. The answer is; **material is just a carrier. What is relevant is "the knowledge and the energy in transforming the material in a form that of use in an intermediate stage or a final stage of the production**". The essence of this statement is that the knowledge & energy really are responsible for added value to the material, not the material content per says.

In Figures 6.1.2, 6.1.3 and 6.1.4, we represent generic activities typical in a manufacturing enterprise. The activities are broken into three considerations:

Figure 6.1.2 "Fixed Assets" covers activities in setting up the manufacturing facility – (per project phases outlined in subsection 5.2).

Fixed Assets

	Task1	L2		K3	E3
	Studies	L		4	2
	Technical	L		4	2
	Permits	L		4	2
DEVELOPMENT	Environmental Study	L		4	2
	Project Schedule	L		4	2
	Project Agreements	L		4	2
	Project Financing	L		4	2
	Engineeering	L		4	2
	Eng'g Deliverables	L		4	2
	Procurement	L		4	2
	Equipment & Bulks	L		4	2
	Manufacturing	L		2	4
EPC	Logistics	L		3	3
	Transport	L		2	4
	Construction	L		3	4
	Buildings & Facilities	L		3	4
	Equpment Installation	L		3	4
	Comissioning & Testing	L		4	4
	COMMERCIAL OPERATION	L		4	4

1) Some tasks are outsourced
2) Labor under K
3) Judges estimates intensity 1 Loweest 4 highest

Figure 6.1.2 Fixed Assets

Fixed Assets deal with the costs of setting up the facilities to start the production operation. They deal with development activities that are needed before the actual implementation of the EPC to complete the facilities with the equipment that is needed. The tasks noted are subject to adjustments depending on the nature and circumstances of the project.

With the facilities ready for operation, we turn to two aspects i) management and corporate support, "Figure 6.1.3"; ii) direct costs for the operation "Figure 6.1.4".

Figure 6.1.3 "Facilities Costs" – these are costs necessary to run the facilities where the production takes place.

	Task1	L2		K3	E3
Facilities	Rent / Equivalent	L			
	Utilities	L		1	3
	Cleaning staff	L		2	2
	Waste Collection	L		2	2
	Maintenace	L		2	2
	Secirity	L		2	3
	Supervision	L		4	2
Mangement & Support	Management	L		4	2
	Corporate	L		3	2
	Centralized Support	L		4	2
	R&D	L		4	3
	Engineering	L		4	2
	Procurement	L		4	2
	Sales & Marketing	L		4	2
	Human Resources	L		4	2
	Tresury	L		4	2
	Legal	L		4	2
	IT	L		4	3
	Communications	L		4	3
	QA/QC	L		4	2
	Safety / Health/ Environmental	L		4	2
	Other	L			

1) Some tasks are outsourced

2) Labor under K

3) Judges estimates intensity 1 Loweest 4 highest

Figure 6.1.3 Facilities costs (at the top) and the management and support / corporate activities (at the bottom).

Figure 6.1.4 production steps are shown.

In Figure 6.1.4 a generic manufacturing including four stages; stage 1 "mining", stage 2 "intermediate". stage 3 "parts manufacturing", stage 4 "assembly". For many enterprises they decide what to outsource and what to keep. Companies evaluate this all the time. According to the strategies that prevail at a given time, the degree of outsourcing is decided.

For smaller manufacturing enterprises, only few tasks are applicable.

		Tasks1		L 2	K3	E3					
Stage 1	Mgmt & Support Facilities	**Minning**	Field		K	E	Mining1	Mining2	Mining3	----	Miningn
		Planning		L	4	2					
		Extraction		L	2	2					
		Sorting		L	3	2					
		Shaping / Forming		L	3	2					
		Inspection / Testing		L	3	2					
		Loading		L	2	2					
		Transport									
Stage 2	Mgmt & Support Facilities	**Intermediate Manufacturing**	Plant		K	L	IntMnfg1	IntMnfg2	IntMnfg3	----	IntMnfgn
		R&D		L	4	2					
		Planning		L	2	2					
		Processing & Shaping		L	3	2					
		Inspection / Testing		L	3	2					
		Loading		L	3	2					
		Transport									
Stage 3	Mgmt & Support Facilities	**Parts Manufacturing**	Factory		K	L	PrtsMnfg1	PrtsMnfg2	PrtsMnfg3	----	PrtsMnfgn
		Machining & Treating		L	4	3					
		Inspection		L	3	2					
		Packaging & Shipping		L	3	2					
		Transport									
Stage 4	Mgmt & Support Facilities	**Assembly Manufacturing**	Plant		K	L	AssMnfg1	AssMnfg2	AssMnfg3	----	AssMnfgn
		R&D		L	4	2					
		Engineering		L	4	2					
		Planning		L	4	2					
		Receiving Parts		L	3	2					
		Subassemblies		L	3	2					
		Final Assembly		L	4	3					
		Inspection & Testing		L	3	3					
		Final Shope Testing		L	4	3					
		Packaging & Shipping		L	2	2					
		Transport									

1) Some tasks are outsourced
2) Labor under K
3) Judges estimates intensity 1 Loweest 4 highest

Figure 6.1.4 Typical activities in a manufacturing plant (four stages shown)

Below are brief discussions of manufacturing of durable product "gas turbine" and "pipe spool", the former more complex than the latter.

Then under nondurable products, "edible oil" extraction is included. The thrust of the presentations is the demonstration of the K-E interplay for the running operations of variety of enterprises. Some simplifying assumptions are noted so enable the reader to keep his /her focus on the K&E model application.

6.1.4.1 Durable Product

a) Gas Turbine

A Gas turbine is a highly engineered product. Let us begin with how an order for a gas turbine to be supplied by company X to an Owner Y, through an EPC Contractor Z who has been contracted by Y to complete a power plant in which the gas turbine will be deployed.

Contractor Z solicits an offer from company X through an RFQ (request for quotation) that spells out the technical requirements and the commercial terms. A proposal manager from company X (usually there are two managers for company X – technical and commercial) takes responsibility for completing the offer, covering the scope of the supply and services, the commercial conditions and of course the price and the payment terms, and including the delivery schedule. The proposal team at company X uses previous records for completed gas turbines of the same type. They introduce appropriate adaptations if need be (to meet the owner's specifications). They check the manufacturing slots to comply with the schedule requirements. Contractor Z reviews the offer and requests clarification. Finally, he makes its recommendations to the owner Y to contract with company X.

If the Owner agrees to the recommendation of the contractor, the contractor negotiates the supply contract with company X and the two sides gets the contract ready to sign. At that time, the owner Y and the supplier X sign the supply contract. The owner Y usually delegates the management of several of its obligations to the contractor Z.

Following each of the signing of the contract completing its actions specified in the condition's precedent to the effective date of the contract, a kick-off meeting is held with the attendance of representatives from the three companies X, Y and Z, triggering the start of the work by company X to fulfill its obligations under the contract.

In the company X, so far, the center of activities is primarily from management and support (refer to Figure 6.1.3). Between knowledge and energy, the weight is much higher on the knowledge side. Involved individuals during this phase usually are highly experienced.

The above is a preface of what happens leading to the order that company X receives. That is the main interest here.

Once the contract is signed, the proposal manager at company X turns over the contract to the contract manager at company X (again two managers: technical and commercial). The contract manager now has the prime responsibility within company X for order #nnnn.

Within company X, the order # nnnn is tracked, meaning all activities under the contract are planned and activities progress is monitored in accordance with the order #nnnn schedule.

The contract manager with his staff identifies the scope items that will be purchased, for instance generator, air intake system, exhaust gas system. Such scope items would be issued to external suppliers or to a sister company/partner to company X in the form of packages. Each package will be tracked separately. The contract team monitors the progress of the suppliers and coordinates with the technical team for attaining approvals of vendor submittals. This is crucial to insure safeguard the project schedule.

For these packages, from the perspective of X the activities are very much knowledge centered, with energy playing a supporting role. On this side, the package suppliers, may use combinations of knowledge and energy in the various activities they undertake. The activities of the suppliers of the packages basically follow the same logic of the manufacturing activities at X but in reduced scales. Within X the management of each of package is assigned to an individual that reports to the contract manager at X.

So far, most of the activities in X on order #nnnn is heavy in knowledge and again centered around activities shown in Figure 6.1.3-- "management and support".

Now we move to focus on the manufacturing of the gas turbine under order #nnnn. The remaining scope--the "core scope"--after taking out the packages noted above consists of these components:

- i) Compressor
- ii) Combustion Chamber
- iii) Turbine
- iv) Control System

Purchase orders for long lead items must include the rotor forging and casings of the compressor and the turbine. We assume that Company X will order these items from suppliers. Leading gas turbine manufacturers have in place frame agreements with their preferred suppliers to speed up the acquisitions of the critical items.

Within X the engineering details are worked out and deliverables that are scheduled for client approvals are forwarded in order not to delay the schedule.

Decisions are made on what is to be sourced out from parts and or subassemblies from items i) – iv) above. Supply contracts are negotiated and signed with suppliers. These contracts must have clear delineation of the limits of responsibility for design, detailing, hold points, and schedule of deliveries.

Again, the above activities within X are centered in knowledge with lesser participation of energy.

The activities at the suppliers follow a parallel pattern as those appearing in Figure 6.1.4. At the suppliers the knowledge and energy contributions vary with each activity.

Now turning to manufacturing within X, with the details of the design and knowing what parts are to be manufactured internally, the parts manufactured proceeds after the material (in whatever form it was specified) arrives to the factory and is stored / distributed to the various machining centers to complete the machining, and the parts inspection per the QA/QC requirements are fulfilled. Also, machining of the lead items bodies takes place. Ready parts / components (such as electrical components for the control system, or valves) are received; they are stored until required for the assembly – stage 4 below.

With the machined main items ready and as well as the other parts, assembly of parts in the main items is completed stage 4 of Figure 6.1.4. The subassemblies are inspected and undergo the required tests. As the subassemblies are ready, the final testing of the integrated assembly of the gas turbine proceeds and the final in shop testing is undertaken. In most cases, owner and contractor witnessing of specified tests take place. With the successful tests, the packaging is completed and the shipments to the site are undertaken.

In the steps outlined above, energy and knowledge take significant effects.

The transport to the construction site usually is subcontracted to transport companies, and in most cases the construction schedule is such that the heavy pieces are placed on the ready foundations upon arrival to the site. Also, the receipt of the ready subcontracted items e.g., the air intake, generator and exhaust system constitute part of the activities that are covered under the gas turbine contract.

The remaining scope of X is to support the contractor staff in the erection and the commissioning at site. Support to the contractor is provided during the final hot commissioning and acceptance test.

b) Pipe Spool Fabrication

Assume that the enterprise is a medium size company with one location, and it is already well established. Thus, it's plant facility is complete and is well equipped. It has the management and support functions as outlined in Figure 6.1.3, Also it has the facility, shop(s) with enough indoor and outdoor spaces. These are equipped with the appropriate cranes and hoists, services such as compressed air as well as furnaces for heat treatment. Forming, bending cutting equipment and machine tools are available with the appropriate specifications. Welding equipment is readily available. The above relates to the fixed assets and now we turn to the specifics for running the operations of the enterprise fabricating pipe spools for various industries including, oil & gas, power, chemicals. The company also serves international projects.

The normal operations start with receiving Request for Quotation to fabricate sets of pipe spools from Contractors. Mostly the initial design of the spools is ready, however, included in the scope is preparation of detailed drawings. The RFQ contains the specifications, the quantities, and the schedule for the deliveries.

The central sales department prepares the offer responding to the technical requirements and the accompanying quotation with the payment schedule and the pertinent commercial terms.

If the Contractor accepts the offer, then the contractor and the Company enter negotiations that lead to signing the contract.

The Company starts the by planning the engineering, the procurement of the material "pipe and fittings" to secure the material deliveries in time, per order # corresponding to the contract. All materials are received inspected / accepted and stored.

With the detailed drawings, the shop initiates the fabrication of the order. This proceeds per a plan for the sets of the pipe spools as issued by the planning department. The fabrication proceeds and the quality inspectors check the spools that are on the shop floor per the quality plans.

The fabricated spools then go to sand blasting, painting, and marking before going to the packaging. In the packaging area the ready spools undergo final inspection and are packaged and readied for shipment.

The above discussions follow the generic manufacturing contents of Figures 6.1.1.- 6.1.4. The intensity of the knowledge and the energy varies from one activity to the next.

6.1.4.1 Non-durable Product

a) Edible Oil Extraction

The company specializes in the production of soybean oil. It has been in this field for decades, with established channels for the supply of soybean and for selling and distribution of its produced oil. It has its own trademark. It has its facility with enough bunker storage for the soybean and the large capacity tanks for produced oil. The transport of the soybean to the factory arrives on trucks and the material handling of the soybean to the presses and the extruders takes place in a mechanized way. All the equipment and the furnaces have been in place for some time and through regular maintenance the equipment is in good condition to support the production. The produced oil is directed to tanks, and it is shipped in tanks to distributers. There is also a bottling unit for oil bottles.

It is obvious that the same generic production activities still represent the manufacturing of the soybean oil. The engineers and the technicians and the equipment operators have the right knowledge. They execute the activities in accordance with the production plans. All quality control and regular testing plus precautions for safety and health are adhered to throughout the factory. The procurement department ensures the timely delivery of the soybean through its sourcing from different locations with different harvesting times. The sales team has developed a good network for marketing and selling the produced oil.

All of the activities, whether directly or indirectly related to the processing, depend on the knowledge that the staff have and the energy (electrical and also oil-fired furnaces) used in the processing equipment. Electrical energy is used to support activities of planning, procurement, etc. In all, it is the combination of knowledge and energy that enter in all the activities of the edible oil enterprise.

6.1.2 Services

In Table 6.1.2.0, data from BEA for the 10 years period between1987 to 1997 are presented for groups of services industries. At the top row of the table the total value of services per year is shown in billions of USD. In the row, the ratio of the total services of that 0f year 1987. One can see almost doubling in 10 years period. Then, in the three parts of the table, groups of services industries that comprise categories in each group are shown. The groups are shown in orange cells. For each group, the top three rows show, total value of the group in billions of USD, ratio of the year total to that of 1987, and the percentage of the group's to the total of the services in each year, respectively. For each category one can see year value in billions of USD.

Table 6.1.2.0 (part 1) USA Total Services per year for the period 1987-1997.

Line		1987	1989	1991	1993	1995	1997
	SERVICES	**3,358,000**	**3,925,000**	**4,183,000**	**4,662,000**	**5,589,000**	**6,370,000**
	Ratio to 1987	1.00	1.17	1.25	1.39	1.66	1.90
34	**Wholesale trade**	286,154	336,422	362,455	403,453	462,210	527,539
	Ratio to item in 1987	1.00	1.18	1.27	1.41	1.62	1.84
	% of Services in the year	8.5%	8.6%	8.7%	8.7%	8.3%	8.3%

35	**Retail trade**	346,064	391,209	407,171	461,887	522,316	579,918
	Ratio to item in 1987	1.00	1.13	1.18	1.33	1.51	1.68
	% of Services in the year	10.3%	10.0%	9.7%	9.9%	9.3%	9.1%
36	Motor vehicle and parts dealers	...	...	...	...	...	...
37	Food and beverage stores	...	...	...	...	...	...
38	General merchandise stores	...	...	...	...	...	...
39	Other retail	...	...	...	...	...	...
40	**Transportation and warehousing**	153,061	168,603	183,874	207,436	229,253	257,255
	Ratio to item in 1987	1.00	1.10	1.20	1.36	1.50	1.68
	% of Services in the year	4.6%	4.3%	4.4%	4.4%	4.1%	4.0%
41	Air transportation	24,324	30,974	32,962	40,434	47,170	54,657
42	Rail transportation	20,883	18,639	20,197	20,333	21,238	19,966
43	Water transportation	4,227	4,715	5,752	5,920	6,075	6,564
44	Truck transportation	44,721	49,079	52,201	59,597	67,389	77,138
45	Transit and ground passenger transportation	7,454	8,316	9,568	10,624	11,986	15,123
46	Pipeline transportation	6,276	5,977	6,068	6,629	6,488	6,953
47	Other transportation and support activities	34,114	38,325	43,530	48,114	50,719	55,862
48	Warehousing and storage	11,062	12,577	13,596	15,785	18,189	20,992
49	**Information**	222,518	253,959	280,954	321,440	356,298	394,074
	Ratio to item in 1987	1.00	1.14	1.26	1.44	1.60	1.77
	% of Services in the year	6.6%	6.5%	6.7%	6.9%	6.4%	6.2%
50	Publishing industries, except internet (includes software)	45,383	53,015	57,119	65,252	73,857	97,171
51	Motion picture and sound recording industries	27,594	33,553	35,156	38,128	39,429	40,289
52	Broadcasting and telecommunications	135,388	150,506	169,593	195,679	216,372	226,340
53	Data processing, internet publishing, and other information services	14,153	16,885	19,087	22,382	26,641	30,275

Table 6.1.2.0 (part 2) USA Total Services per year for the period 1987-1997.

Line		1987	1989	1991	1993	1995	1997
	SERVICES	3,358,000	3,925,000	4,183,000	4,662,000	5,589,000	6,370,000
	Ratio to 1987	1.00	1.17	1.25	1.39	1.66	1.90
54	**Finance, insurance, real estate, rental, and leasing**	842,066	972,751	1,097,343	1,248,521	1,406,525	1,612,356
	Ratio to item in 1987	1.00	1.16	1.30	1.48	1.67	1.91
	% of Services in the year	25.1%	24.8%	26.2%	26.8%	25.2%	25.3%
55	Finance and insurance	276,796	317,288	375,657	433,448	484,994	577,187
56	Federal Reserve banks, credit intermediation, and related activities	146,235	153,953	183,507	202,414	208,586	233,146
57	Securities, commodity contracts, and investments	41,379	45,722	46,840	64,791	75,544	121,296
58	Insurance carriers and related activities	81,795	110,942	137,485	155,837	189,711	215,595
59	Funds, trusts, and other financial vehicles	7,386	6,671	7,826	10,406	11,154	7,150
60	**Real estate and rental and leasing**	565,271	655,463	721,686	815,073	921,531	1,035,169
	Ratio to item in 1987	1.00	1.16	1.28	1.44	1.63	1.83
	% of Services in the year	16.8%	16.7%	17.3%	17.5%	16.5%	16.3%
61	Real estate	513,205	590,972	653,071	736,673	832,008	929,207
62	Housing	...	...	...	...	...	...
63	Other real estate	...	...	...	...	...	...
64	Rental and leasing services and lessors of intangible assets	52,066	64,491	68,615	78,400	89,522	105,961
65	Professional and business services	388,621	476,992	531,627	607,964	690,022	840,553
	Ratio to item in 1987	1.00	1.23	1.37	1.56	1.78	2.16
	% of Services in the year	11.6%	12.2%	12.7%	13.0%	12.3%	13.2%
66	Professional, scientific, and technical services	223,803	288,763	326,292	370,261	412,684	499,682
67	Legal services	59,671	76,164	85,861	94,263	95,535	103,987
68	Computer systems design and related services	21,672	27,836	33,304	41,081	51,004	75,488
69	Miscellaneous professional, scientific, and technical services	142,460	184,763	207,128	234,918	266,145	320,207
70	Management of companies and enterprises	77,200	78,330	81,991	93,132	103,788	125,467

71	Administrative and waste management services	87,617	109,898	123,344	144,570	173,550	215,405
72	Administrative and support services	76,504	96,019	108,160	126,564	153,151	193,546
73	Waste management and remediation services	11,113	13,880	15,184	18,007	20,399	21,859

Table 6.1.2.0 (part 3) USA Total Services per year for the period 1987-1997.

Line	Line	Line	Line	Line	Line	Line	Line
	SERVICES	**3,358,000**	**3,925,000**	**4,183,000**	**4,662,000**	**5,589,000**	**6,370,000**
	Ratio to 1987	1.00	1.17	1.25	1.39	1.66	1.90
74	**Educational services, health care, and social assistance**	278,904	339,199	414,221	484,559	538,460	590,585
	Ratio to item in 1987	1.00	1.22	1.49	1.74	1.93	2.12
	% of Services in the year	8.3%	8.6%	9.9%	10.4%	9.6%	9.3%
75	Educational services	33,600	38,984	45,642	53,657	65,184	77,078
76	Health care and social assistance	245,304	300,214	368,579	430,902	473,276	513,507
77	Ambulatory health care services	120,795	148,716	182,746	213,153	235,377	250,899
78	Hospitals and nursing and residential care facilities	108,575	131,619	162,165	188,728	204,491	223,126
79	Hospitals	...	...	...	...	...	...
80	Nursing and residential care facilities	...	...	...	...	...	...
81	Social assistance	15,935	19,879	23,667	29,021	33,407	39,483
82	**Arts, entertainment, recreation, accommodation, and food services**	153,369	185,095	206,565	232,259	257,310	301,782
	Ratio to item in 1987	1.00	1.21	1.35	1.51	1.68	1.97
	% of Services in the year	4.6%	4.7%	4.9%	5.0%	4.6%	4.7%
83	Arts, entertainment, and recreation	33,607	42,042	50,423	59,145	66,657	80,141
84	Performing arts, spectator sports, museums, and related activities	16,137	19,274	24,040	28,526	31,732	37,156
85	Amusements, gambling, and recreation industries	17,470	22,769	26,383	30,619	34,924	42,984
86	Accommodation and food services	119,762	143,053	156,142	173,115	190,654	221,642
87	Accommodation	36,178	41,901	46,059	52,053	59,544	70,738

88	Food services and drinking places	83,584	101,152	110,083	121,062	131,110	150,903
89	**Other services, except government**	121,280	145,014	155,781	179,419	204,834	230,311
	Ratio to item in 1987	1.00	1.20	1.28	1.48	1.69	1.90
	% of Services in the year	3.6%	3.7%	3.7%	3.8%	3.7%	3.6%

It is interesting to note the growth of the services versus the manufacturing in the economy of the USA, this is depicted using the totals of the manufacturing from Table 6.1.1.o and the totals of the services from Table 6.1.2.0 for years 1987 and 1997, please refer to Table 6.1.2.0 (part 4)

Table 6.1.2.0 (part 4) Comparison of the splits of manufacturing and services for 1987 & 1997

Year	**1987**		**1997**	
	billion $	% of sum	**billion $**	% of sum
Manufacturing	878.80	21%	1,382.90	18%
Services	3,358.00	79%	6,370.00	82%
SUM	4,236.80		7,752.90	

In Figure 6.1.2.1 (part 1) typical activities common to all service groups apply. In Figures 6.1.2.1 (part 2) and (part 3) various groups of service industries are shown.

					K	E
i)	**Facilities**					
	Rent / Equivalent		L			
	Utilities		L		1	3
	Cleaning staff		L		2	2
	Waste Collection		L		2	2
	Maintenace		L		2	2
	Secirity		L		2	3
	Supervision		L		4	2

					K	E
	Mgmt & Support					
	Corporate		L		3	2
	Centralized Support		L		4	2
	R&D		L		4	3
	Engineering		L		4	2
	Procurement		L		4	2
	Sales & Marketing		L		4	2
ii)	**Human Resources**				4	2
	Tresury		L		4	2
	Legal		L		4	2
	IT		L		4	3
	Communications		L		4	3
	QA/QC		L		4	2
	Safety / Health/ Environmental		L		4	2
	Other		L			

Figure 6.1.2.1 (part 1): Services Activities " I) Facilities – II) Management Support"

				K	E
III.1	Wholesale trade				
III.2	Retail trade				
	Motor vehicle and parts dealers				
	Food and beverage stores				
	General merchandise stores				
III.3	**Transportation and warehousing**				
	Air transportation				
	Rail transportation				
	Water transportation				
	Truck transportation				
	Transit and ground passenger				
	Pipeline transportation				
	Other transportation and support				
	Warehousing and storage				
III.4	**Information**				
	Publishing industries, except				
	Motion picture and sound recording industries				
	Broadcasting and telecommunications				
	Data processing, internet publishing, and other information services				
III.5	**Finance and insurance**				
	Federal Reserve banks, credit intermediation, and related activities				
	Securities, commodity contracts, and investments				
	Insurance carriers and related activities				
	Funds, trusts, and other financial vehicles				
III.6	**Real estate and rental and leasing**				
	Real estate				
	Housing				
	Other real estate				
	Rental and leasing services and lessors of intangible assets				

Figure 6.1.2.1 (part 2) Different Services Activities " III) CORE Cont'd

					K	E
III.7	**Professionalscientific, and technical services**					
	Legal services	L	L			
	Computer systems design and related services	L	L			
	Miscellaneous professional scientific and technical services	L	L			
III.8	**Administrative and waste management services**					
	Administrative and support services	L	L			
	Waste management and remediation services	L	L			
III.9	**Educational services**	L	L			
III.10	**Health care and social assistance**					
	Ambulatory health care services		L			
	Hospital and nursing, residential care facilities		L			
	- Hospitals		L			
	- Nursing and residential care		L			
	- Social assistance		L			
III.11	**Arts, entertainment and recreation**					
	Performing arts, spectator sports, meuseums and related services		L			
	Amusments, gambling and recreation industries		L			
III.13	**Accommodation and food services**					
	Accommodation		L			
	Food services and drinking places		L			

Figure 6.1.2.1 (part 3) Different Services Activities " III) CORE Cont'd"

6.1.2.1 Example #1 Education - Private school

The focus is on the operation stage of a private school (please refer to subsection 5.1); it is assumed that the fixed asset of the school is in place.

The principal is to manage the operation of the school for the two semesters per the approved annual budget. He is to make sure the teachers, instructors, and support-staff are in place. The administrator together with her staff is managing new admissions and registration for the classes. The classrooms are readied for the beginning of the school year and all the teachers are briefed with schedule and other relevant administrative policy / procedures.

The teachers take the main tasks in delivering in the classes using different means for effective learning of the students. The students also receive guidance / classes in sports and extra curriculum. Figure 6.1.2.1 depicts the interplay of the knowledge and energy in the operation of the school.

6.1.2.2 Example #2 Medical - Family Doctor

A family doctor has set up his clinic, the fixed assets are in place with all the equipment he needs installed and properly maintained through maintenance contracts with the original suppliers. He has retained an experienced nurse that supports him in receiving patients, scheduling the visits of the patients, and helping him with a number of tasks.

Daily, the clinic receives several patients, and the patients are examined/diagnosed. The doctor refers the patient that needs medically specialized examinations to laboratories and hospitals, so he can complete the investigations for patients.

The payments of the patients including the insurance paperwork is completed by the patient and administered by the nurse; an outsourced accountant handles the follow up.

Again, the interplay of the knowledge--the doctor, the nurse, and the outsourced services with the energy consumed (computers, medical equipment, etc.) in performing all the activities is clear.

6.1.2.3 Example #3 Transport -Taxi service

A small enterprise owned by a proprietor has a small fleet of 10 caps. The ten vehicles have been owned by the enterprise and they have maintenance contracts with the car dealer from which the cars were bought. The enterprise has a dispatch center where an operator (3 shifts) mans the office 24/7. Beside the dispatch handling the lead dispatcher who has had seniority in the company also, handles administrative tasks pertinent to fueling the caps with the cap drivers. The proprietor uses a small building conveniently located with the dispatch center and the garage for the cars when not in use. He himself administers the scheduling of the drivers on the 10 vehicles. He also receives from the drivers the fares they collected.

The activities described above follow the generic presentation of the interplay of knowledge (of the proprietor, the dispatchers, the drivers) and the energy of consumed in the office equipment and the cars.

6.1.3 Construction

6.1.3.1 Example #1 Solar PV Plant

A general contractor that specializes in EPC for solar PV plants receives a request for proposal RFP (Request For Proposal) from a developer for a 50 MW (Photo Voltaic) PV project.

A proposal manager is assigned to lead the team to prepare the proposal. The team of engineers studies the technical specifications and the requirements in the RFP. The proposal manager submits requests for clarifications to the developer and in parallel the team identifies the required equipment and bulk material and prepares request for quotations for these. The civil staff prepares the design, drawings of the plant based on the site data in the RFP. The technical team prepares the bills of material (BOM) for the construction. The procurement department issues the RFQs (Request for Quotation) to obtain quotations from suppliers. The received technical parts of the offers from the suppliers are evaluated. Technical clarifications to suppliers are issued as necessary. Then the technically accepted offers are evaluated commercially to establish the selected suppliers. The construction staff prepares the pricing for the construction based on the BOM (Bill of Material). The proposal team completes other parts for the project execution in response to the RFP requirements. The offer, together with the bid bond, is submitted to the developer by the due date.

The proposal team receives requests for clarifications from the developer and the developer completes its assessment and decides that the contractor is shortlisted and invites the contractor for negotiations of the EPC (Engineering Procurement Construction) contract. The negotiation goes smoothly, and the EPC contract is signed.

Through the above steps with the contractor the activities are knowledge-centered (of the contractor proposal team) with the energy contributions of all the tools the team uses e.g., computers, communications, etc.

With the signed contract in hand, a project manager is assigned, and he/she takes charge. The project team is formed. The engineering team handles all the engineering deliverables including detailed RFQs for the supplies and bulks. The engineering team prepares the detailed drawings for the various disciplines (civil, electrical, and mechanical) encompassing the plant facilities and systems. The procurement department proceeds with getting quotes for the required supplies (equipment and bulks) and for services including logistics for marine and inland transport, site facilities for the construction team. Detailed plan for the construction is prepared factoring the delivery dates of the main equipment, in an optimized way.

The mobilization of the construction team under the supervision of the site superintendent is initiated. The civil group proceeds with site clearing and setting up the internal roads for the other disciplines to commence the erections. The civil team completes the control building. Civil also prepares the cable trenches for the DC (Direct Current), AC (Alternate Current) cable, fire, and power cables for the security system. The mechanical and electrical discipline supervisors with their crews complete the mechanical (mounting structures, placement of panels) and electrical follow up with DC cabling and inverters / transformers installation and connections. Then civil and electrical lay the AC cables. The electrical group completes the installation of the control building equipment. In parallel, the QA/QC and safety officers oversee the quality and health, safety, and environmental requirements.

The construction activities involve the knowledge of supervisors, skilled technicians and the labor force aided by the energy of all the equipment at the site--cranes, vehicles, power tools, etc.

Upon the completion of the construction, cold commissioning is initiated and inspections and tests for subsystems are undertaken. Once the grid connection is achieved, then hot commissioning proceeds ending with plant performance tests. The activities of the commissioning take place under the overall responsibility of the commissioning manager with participation of supplier's representatives for key equipment. The commissioning period is marked by large consumption of electrical energy in the debugging of the equipment.

From the above we see in this example of a construction project, the close interplay of knowledge and energy throughout the stages of the project.

6.2 Historical Data

All enterprises keep records, all departments and sections maintain retrievable files that document the evolutions of the enterprise. For example, at R&D department, records of the trials are to be kept, both the successful and the failed. There could be benefit at future time from a trial that was abandoned in the past for one reason or another. At the engineering department, all the details of engineering documents, e.g. design bases, calculations, drawings, etc. of the designs of parts and subassemblies must be archived in good order for ease of retrieval. The basis of design is kept. All past experiences could be quite useful in introducing improvement and new products. The same applies also for services. As enterprises strive towards continuous improvement, they make use of where they are at and how they got to the current point as they try to optimize the production for the future.

6.3 Planning & Management

In operations we are always concerned with the three main parameters: schedule, cost and quality. This is always the case whether the enterprise is in manufacturing or in services, or mixed enterprise (e.g., construction). Going back to the simplistic process representation in Figure 1.3.1. Planners use the terms "breakdown structure, tasks, subtasks" and the structure can grow depending on the size of a project and complexity. If we refer to Figure 1.3.1 when we have processes that go in a successive manner, then we have the term "precedent", meaning that a process cannot start except once another process has been completed. In planning when we talk of schedule, where we are interested in the overall timetable, that is for say a construction project from start to finish, and for manufacturing of a product, the overall duration to produce one or sets (groups) of products. For services organizations we still need to deal with planning and the time of executions of the various processes.

From the perspective of the K&E model, costs would be those associated with the K&E inputs (please refer to subsection 5.5). The discussions in subsection 5.7 talk of the knowledge & energy that had entered (become embedded) in various means that are deployed or used in the execution of the process. Furthermore, the overall integration of the processes entails additional knowledge and energy inputs.

A comment is warranted: in the cost systems now in use in enterprises, we generally talk of direct and indirect costs.

Further quality comes in making sure that the input to a process is in accordance with specifications, and further that the execution is also per the requirements and specifications of each process. Collectively, when each process is successfully completed, the ultimate result is in accordance with the perceived objectives.

Without any doubt, management has a very important role in guiding the planning, and then in monitoring the performance by the execution teams. This is heavily dominated by the knowledge of the various management levels.

6.4 Additional Costs

In the K&E interplay premise introduced in section 5, the primary factors controlling products and services are knowledge and energy. In subsection 6.1 the focus was on the running operations of the enterprise. These included management and support functions as well as facilities related activities. From the perspective of the conventional KLEMS the costs associated with the K (please refer to

Figure 6.1.3) come from the allocated fixed costs. These costs relate to phases 1&2 of the project that was undertaken to build the facilities and set up the operations--the initial capital invested (refer to the discussions of subsection 5.2). The investments are further elaborated on in section 8.

In the following subsection, additional costs come up due to government actions; for instance, reporting on environmental issues, renewing licenses, filing and paying taxes.

Even the activities undertaken by the staff of the enterprise that work on the governmental issues perform those activities with the combination of knowledge and energy.

6.5 Government

As discussed in subsection 6.3, the enterprise needs to deal with many different governmental agencies. These involve interactions between the enterprise and the various concerned agencies, e.g. submittal applications, updating, providing clarifications, paying fees, and getting permits, licenses.

The government agencies have many services that directly or indirectly affect the operations of the enterprises. Providing the activities behind these services involve knowledge and energy.

It is worth examining the wide spectrum of activities the government handles. Considering the constitution, the overall structure of the government has three branches--executive, legislative, and judicial. For instance, in the USA, the executive branch is headed by the president, with various departments headed by the secretaries that are nominated by the president and confirmed by the houses of representatives & senate (legislative branch). In the USA, the federal government follows this set up. I deleted the next section about the states. I don't think it's a very useful example.

From the perspective of knowledge, we see the importance that the government plays in terms of nourishing knowledge.

Starting in education at all levels, concerned government agencies aim to develop the capabilities and skills of the students and instill the foundation of knowledge in the growing part of the population before they join the work force. The educational agencies ensure uniformity of curriculums while giving flexibilities to school administrators and teachers to attain optimum results. The concerned agencies distribute the funding to public schools. They also have oversight on the operations of public and private schools. All involved in the educational activities are making use of knowledge to perform the activities to make the schools work properly; the teachers and the staff of schools and colleges depend on their knowledge to *train the students. The producing of the books is based on knowledge. The forgoing tasks are accomplished* through energy.

Government sponsors research to induce new developments including basic research through grants to universities. Research institutes play a significant role in developing knowledge in various fields. Government agencies support setting up of incubators of new applications. Research leads to innovations, a key to the optimizations of knowledge and energy towards reaching best results for products and services. Knowledge is enriched – new additions to knowledge associated with this energy is consumed.

Government agencies administer registration of patents. They also play an important role in protecting the intellectual rights of the inventors and the authors. This is necessary for the flourishing of ideas and knowledge sustainability.

On the energy side government regulates the electrical industry. It needs to secure the availability of fuel for the fossil plants (coal, oil and natural gas), nuclear and renewable energy. Government agencies consider applications of licenses for new power plants. Government agencies grant licenses to TSO's and DSO's. Agencies ensure that operations are within the licensed environmental permits for all assets in the electrical system. Thus, the government has a vital role in the electrical energy supply.

Similarly, for transport, the government agencies ensure sufficient supplies of fossil fuel – for cars, trucks, buses, and electrical energy for e-mobility. Governmental agencies plan, sponsor, and oversee infrastructure of the roads and highways. They do similar activities for railroads. Also, they plan and sponsor building ports and airports.

Government agencies negotiate trade agreements with other countries resulting in removal of barriers and significant effects on imports and exports. This extends to entering in cooperation agreements with other countries for mutual benefits.

Security of the country is maintained through coordination of the external affairs (state department) and the various military forces, national guards. Police effort to maintain secure neighborhoods is part of government activities.

The legal branch of government is responsible for administration of laws passed by the legislative branch hence resolving conflicts and in cases of crimes applying the appropriate penalties.

Government agencies regulate the health industry, the pharmaceutical industry and hence affect the well- being of the individuals of the society.

As will be discussed in section 7 and 8, government has significant influence on the availability of money and the interest rate affecting savings and loans. Also, the government regulates and has oversight over the banking industry.

The purpose of the above discussion is not to cover all activities of the government but to address examples – in all activities, knowledge is key and the support of energy to make the activities be realized is clear.

6.6 Society – Customer Attitudes

In the previous subsections we applied the qualitative model of the K&E to several manufacturing and service enterprises. We also addressed government activities again underling the K&E.

We would like to supplement this by stressing the importance of customer attitudes, since at the end enterprises provide their products or services to customers that are interested and able to afford the price of the offered product / service. The interest depends on reaching the customer and convincing the customer of the merits. In this respect, the enterprises could be initiators to spark customer interest in a new or a refined product / service. Enterprises also listen to customers and seek their feedback and monitor the changes of the attitudes of customers.

6.6.1 Business to Business

A manufacturing enterprise dealing with other manufacturing enterprises on its supply chain needs to have a good oversight on various aspects relating to the quality and timeliness of deliveries. On the other side the enterprise needs to keep an eye on changes that affect its customers. This means watching changes in the market, changes that affect the suppliers and the customers. The enterprise must watch the competition affecting its suppliers and customers. This requires activities that involve knowledge and energy to secure sound information on which the right strategies are put in action. The sought information covers wide ranges--financial health of the company, workload, changes in the organization, technical specifications changes, applicable governmental decrees.

6.6.2 End Customer

Knowing the changes of society / customs - what people are buying - is very important to the enterprise. The question of who sets the trends--the enterprise or the customer – the reality it is both have contributions in setting new trends. In recent years, large companies have emerged as trendsetters for many products and in deliveries of services. We can see this in smart phones or fashion designers. The lifetime of a product is an example of a characteristic that the producer needs to investigate continuously to anticipate the mood of the customer.

As will be addressed in section 7, disposable income affects the spending of the customer. Companies target market shares in specific segments.

6.7 Recent Achievements & Ongoing Digitization (Examples)

6.7.1 Recent Achievements

6.7.1.1 Example #1 – Gas Turbine Efficiency

Over the last 40 years gas turbine efficiency saw a doubling in the efficiency from 20% to 40%. In a combined cycle configuration, we see an increase from 30% to 60%. This means that for the same amount of electrical power, we now burn half the amount of fuel compared to what we would get in gas turbine of the late 1980s era. Thanks to improvements in materials, design details, and manufacturing process, higher combustion temperature has been reached leading to the significant improvement in the performance (please refer to section 2).

6.7.1.2 Example #2 – Electrical Energy Consumption in Aluminum manufacturing

Manufacturing of aluminum use electrolysis of bauxite ore (Hall Heroult process) consuming large amounts of electrical energy. Thanks to improvements in the technology, the energy consumption per ton of aluminum went down from 18 -17 MWh to 14-13 MWh.

6.7.1.3 Example #3 – Electrical Energy Consumption in PV panels

A good measure of the performance of PV (Photo Voltaic) panels is how long it will take; for the energy produced from the panel to equal the energy that went into its manufacturing of the panel. Ten years ago, it was estimated as 3-4 years, for currently produced PV panels, 1 to 1.2 years. This is a significant improvement given that lifetime for PV panels of 2020 is 20 -30 years.

6.7.2. Ongoing Digitization

A manifestation of knowledge that has become a major propeller in the various sectors is "digitization", some people regard as the fourth revolution for mankind.

Referring to section 4, we discussed different important technologies (that fall under knowledge) which have found their way in wider applications in many fields. At the center of this is data collection with efficient retrieval and artificial intelligence. We shall discuss some hereafter:

6.7.2.1 Electric Grid

With the increased number of renewable plants in both the TSO and DSO of the electric grids both in the US and Europe, the TSO and DSO has seen a major change. Only a few years ago the TSO and DSO received generated electricity from a small number of power plants – they generally had large capacity with well-known characteristics that enabled the system operator to dispatch the units from the different generating plants in an order of merit (old regime discussed in section 2). Even when distributed CHP plants were introduced, the dispatch of power plants in the generation was straightforward. However, with solar and wind plants – with varied capacities from small MW to large 500 MW, and intermittent production depending on the weather and time of day, and with these units connected directly to either high or intermediate voltage of transmission or the distribution parts of the grid, now controlling the operation of the grid with all the electrical inputs versus the loads of the customers represents major challenge compared to the old regime. Smart meters (thousands) have now been installed/more being added in different nodes connected to the generator / user and through these meters the operator(s) of the transmission / distribution would be able to monitor and control and manage the operation of the grid always ensuring the required reliability and quality of the electricity to the customers.

AI is being applied towards gaining insights from data collected from the smart meters. This is a major endeavor that many are contributing to and we hope for getting good results in the near future.

6.7..2.2 PV plants SCADA

PV plants are composed of thousands of PV panels interconnected to inverters (in the case of string inverters hundreds) with combiner boxes, recognizing where any fault is quickly is essential for achieving the performance of the plant according to the design by identifying and fixing the problem in a short time. For these manufacturers have provided compatible means to enable the SCADA (Supervisory Control Architecture and Data Acquisition) / monitoring system to identify and isolate the problem for quick remedial action.

6.7.2.3 E-Home (Household appliances)

Through communications wireless and cables, it is now possible to monitor the working of the AC and refrigerator, oven, washing machine, dishwasher, TV and lights. It is possible to schedule the times of On - Off. This of course helps houses with two working partners and even for the non-working wife to optimize the operations of the appliances with obvious savings in energy. Much of technology depends on smart meters and makes use of signals transmitted to monitors and to smart phones. The obtained data enables getting a good view with details on electricity consumption. The data in turn can be analyzed and appropriate trends can be deduced.

6.7.2.4 E-Mobility

Electric cars fitted with sensors and computers are found in the market with ever increasing technology (knowledge) and electric engine (energy) with optimization of battery storage.

6.7.2.5 E-Health Services

The deployment of AI in health services is seeing a lot of interest from the medical profession and through the increased collaboration between the health specialists and the AI professionals many applications have been developed and many more are under development.

6.7.2.6 E-banking

The banks are also deploying with the aid of the AI professionals many applications with the objective of simplifying the various transactions to the clients to accomplish through reliable means apps on the smart phones. The potential benefits to business and individual consumers are enormous.

6.7.2.7 Two Additional specific applications:

c) Drone use in Construction Monitoring

In recent years, drones have been deployed to monitor construction progress. With the appropriate communication system on the ground, the drone communicates the photos it takes which are in turn transmitted to the office for analysis and reports production daily, providing a tool for headquarters to get firsthand information on the status that complements the reports of the construction manager.

d) Robots Cleaning PV modules

In recent years the manual and brush machines cleaning of the PV panels saw robot-cleaning machines taking a big part of the PV panel cleaning market.

6.8 Knowledge Prominence

As we can see from subsection 6.1, services now represent the bigger share contribution of private enterprises to the GDP, more than 70-75 % services to 30 -25% for manufacturing with continuous increase in favor of services. For services, knowledge mostly has the bigger share of inputs between the K&E. We can see the increasing emphasis on knowledge in the compositions of work force in various enterprises.

Recently, we could see the high valuations of stocks of tech companies. We see that oil companies stock values have dropped considerably from their highs in the later part of the last century. Many reasons could be given, among which is the large drop in oil prices. And now societies favor renewable energy because of the major concern of global warming associated with gas emissions (combustion products). We also see that energy conservation programs (knowledge based) have produced positive results in the US and Europe. Please refer to decrease of electricity consumption in recent years (subsection 7.3).

A remark that must be added is that oil majors still possess valuable technologies needed to get oil from the ground to the right form needed for operating a plane/vehicle, or for a power plant generating electricity.

As is discussed in subsection 7.1, the supply in relation to the declining demand is behind the drop of fossil fuel prices that we are seeing in recent years. The downturn in economies from March 2020 because of the COVID 19 pandemic of course added to the decrease of the fossil fuel demand.

6.9 References & Bibliography

1. **Safwat, H.H.,** "The Project's Manager & Team", presented in the 4th International Symposium – Project Management for Development & Growth, Cairo-Egypt, February 7-10, 1993
2. **Safwat, H.H.,** and **Higgy, H.R.,** "Technology Transfer – The Partnering Process", conference Proceedings of the 4th International Conference on Management of Technology, Miami, Florida, February 27 – March 4, 1994.
3. **Safwat, H.H**, "On the Integration of Automation Technology Staff in a Project Team", 6th International Forum on Technology Management, organized by Jupiter Consortium, Amsterdam, The Netherlands, October 1996.
4. **Safwat, H. H.,** "Technology Transfer; The Consultant's Role", the Conference Proceedings of the 11th International Conference on Management of Technology, Miami, Florida, March 10-15, 2002.

5 **U.S. Bureau of Economic Analysis** - https://www.bea.gov/data

In Part II the knowledge & energy interplay model was introduced (section 5).
In section 6, the application of the K&E model to several enterprises was presented.

PART III
MACROECONOMICS AND INVESTMENTS

In Part III, the second word in the book title "economics" is presented through two sections, section 7 "macroeconomics" followed by "investments" in section 8. The emphasis is on covering material from sources of information in these two fields in preparation for the introduction of the quantitative model of the K&E included in section 9 of part IV.

Section Seven: MACRO ECONOMICS

The author at the outset does not claim to be an authority in economics but in this section, he has tried to include some of the information he gathered that help in demonstrating the dependence of economy on the two cornerstones, knowledge and energy. Unfortunately, the author misses Professor Oweiss who was a key participant in the preparation of the Reference. Economics publications are rich with theories and opinions of respected scholars the interested reader can consult. Economics is a branch of social science focused on the production, distribution, and consumption of goods and services. In sections 5 and 6 we covered how knowledge and energy are the cornerstones for all the activities of enterprises. This was highlighted in section 6 through applying of the K&E interplay model to manufacturing and services enterprises.

In subsection 7.1 we start by introducing the well-known principle of supply and demand.

Then we turn to macroeconomics to discuss the economy of a nation. The US being the leading economy of the world has been chosen. Economic data of the US (with related energy and labor) are presented in subsections 7.2-7.4. In some instances, data / comments data from other countries are included. Later several subsections cover important topics of economics. The intention is to present material generally found in the economics domain to underline the relation / dependence of the fundamentals of these topics to knowledge and energy.

7.1 Aggregate Supply & Demand

The consumer is the propeller of the economy. The consumer purchases of a certain product or service are affected by the supply of the product /availability of the service. In the meantime, effect of the demand on the price is well known. This leads to the fact "supply and demand go in tandem". It is hard to know which one comes first. The fact is supply and demand control the sale/consumption of any product or service. That is how market prices are set.

First the consumer should have an interest in a specific product or service (to satisfy a need or a want). Secondly the potential customer must have money or part of his/her income or savings money to buy the product/ get the service. The consumer decides according to his/her priorities to take steps toward making the purchase, this is when supply & demand kick in with the known (advertised) market price.

Today knowledge plays a very important role in the economy as a major influencer in many ways. Through technologies and innovations, new products and services or improved products and services become available on the market. Through different means, businesses publicize and market their products and services.

7.2 Economic Data

A national economy is the result of a set of processes that involves the following main factor: nation's culture, values, education, technological evolution, history, social organization, political structure, and legal systems, as well as its geography, natural resource endowment, and ecology. The economy functions through the conditions of these factors.

In macroeconomics, the economy of a nation, indices are used and are produced / published periodically in reports by government agencies. In the US, the Bureau of Economic Analysis BEA publishes data for economic indices. On the level of the US, Key Indices are discussed below with references to extracts from the BEA data publications.

7.2.1 GDP

For the USA, GDP "Gross Domestic Product" is a comprehensive measure of the U.S. economy and its growth. Please refer to Table 7.2.1 below for the 2017, 2018 and 2019 GDP.

Table 7.2.1

RELATION OF GROSS DOMESTIC PRODUCT, GROSS NATIONAL PRODUCT, AND NATIONAL INCOME
[Billions of dollars]

Line		2017	2018	2019
1	**Gross domestic product (GDP)**	**19,543.0**	**20,611.9**	**21,433.2**
2	Plus: Income receipts from the rest of the world	1,032.7	1,142.9	1,169.8
3	Less: Income payments to the rest of the world	740.4	858.2	900.2
4	**Equals: Gross national product**	**19,835.3**	**20,896.6**	**21,702.9**
5	Less: Consumption of fixed capital	3,112.9	3,265.0	3,420.9
6	Less: Statistical discrepancy	-131.4	-58.0	12.8

7	**Equals: National income**	**16,853.9**	**17,689.6**	**18,269.1**
8	Compensation of employees	10,422.5	10,950.1	11,432.4
9	Wages and salaries	8,471.5	8,894.2	9,309.3
10	Supplements to wages and salaries	1,951.1	2,055.9	2,123.1
11	Proprietors' income with inventory valuation and capital consumption adjustments	1,509.0	1,585.9	1,657.7
12	Rental income of persons with capital consumption adjustment	721.9	759.3	787.1
13	Corporate profits with inventory valuation and capital consumption adjustments	2,114.5	2,243.0	2,250.5
14	Net interest and miscellaneous payments	636.4	619.1	573.4
15	Taxes on production and imports less subsidies	1,304.1	1,381.5	1,417.5
16	Business current transfer payments (net)	150.7	156.6	158.0
17	Current surplus of government enterprises	-5.3	-5.8	-7.6
	Addenda:			
18	Gross domestic income (GDI)	19,674.4	20,669.9	21,420.4
19	Average of GDP and GDI	19,608.7	20,640.9	21,426.8
20	Statistical discrepancy as a percentage of GDP	-0.7	-0.3	0.1

Table 7.2.1 above shows the US **GDP** in billion USD for the period 2017 to 2019. It also shows the **GNP** Gross National Product (GDP with adjustments "plus Income receipts from the rest of the world, less: Income payments to the rest of the world"). Table 7.2.1 also shows the **National Income** (calculated from GNP "less: consumption of fixed capital, less: statistical discrepancy").

7.2.2 GDP Growth Rate

Table 7.2.2 Growth Rate_Real Gross Domestic Product 2017-2019

REAL[1] GROSS DOMESTIC PRODUCT AND RELATED AGGREGATES: PERCENT CHANGE FROM PRECEDING PERIOD AND CONTRIBUTIONS TO PERCENT CHANGE				
Line		2017	2018	2019
	Percent change from preceding period			
	Gross domestic product (GDP) and related aggregates:			
1	GDP	2.3	3.0	2.2
2	Goods	3.5	5.7	4.1
3	Services	1.7	2.0	1.6
4	Structures	3.2	1.4	-0.4
5	Motor vehicle output	-0.8	7.5	0.3
6	GDP excluding motor vehicle output	2.4	2.9	2.2
7	Nonfarm business gross value added[1]	2.8	3.5	2.5
	Contributions to percent change in real gross domestic product			
	Percent change at annual rate:			
8	**Gross domestic product**	**2.3**	**3.0**	**2.2**
	Percentage points at annual rates:			
9	Goods	1.02	1.66	1.19
10	Services	1.04	1.22	1.00
11	Structures	0.27	0.12	-0.03
12	Motor vehicle output	-0.02	0.21	0.01

1. Consists of GDP less gross value added of farm, of households and institutions, and of general government.
Source: U.S. Bureau of Economic Analysis

Table 7.2.2 gives the indication of the economy growth rate

7.2.3 GDP Historical Data (1990-2020)

The quarterly data of GDP was extracted and using an average for each year from the four quarters, the average GDP was obtained and is plotted in the Figure below.

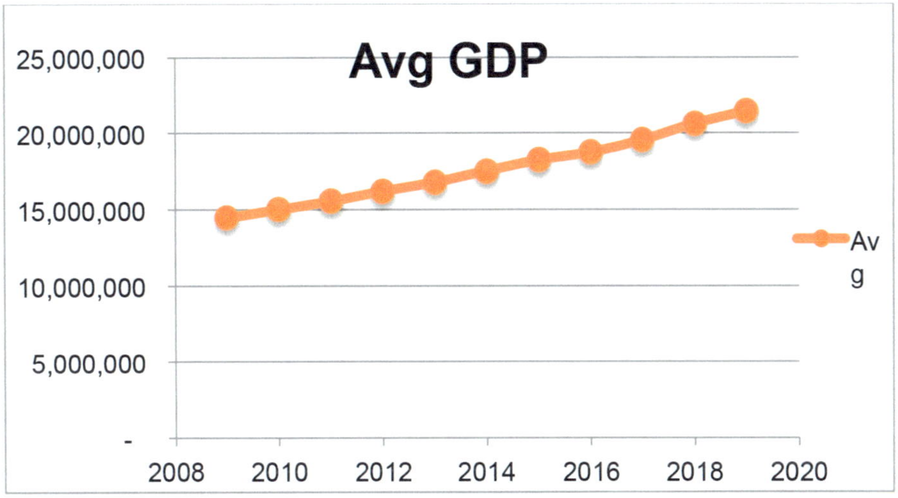

Figure 7.2.3.1 Plot of GDP change between 2009 -2019 (Derived From Federal Reserve Board Data

7.2.4 GDP Growth Rate

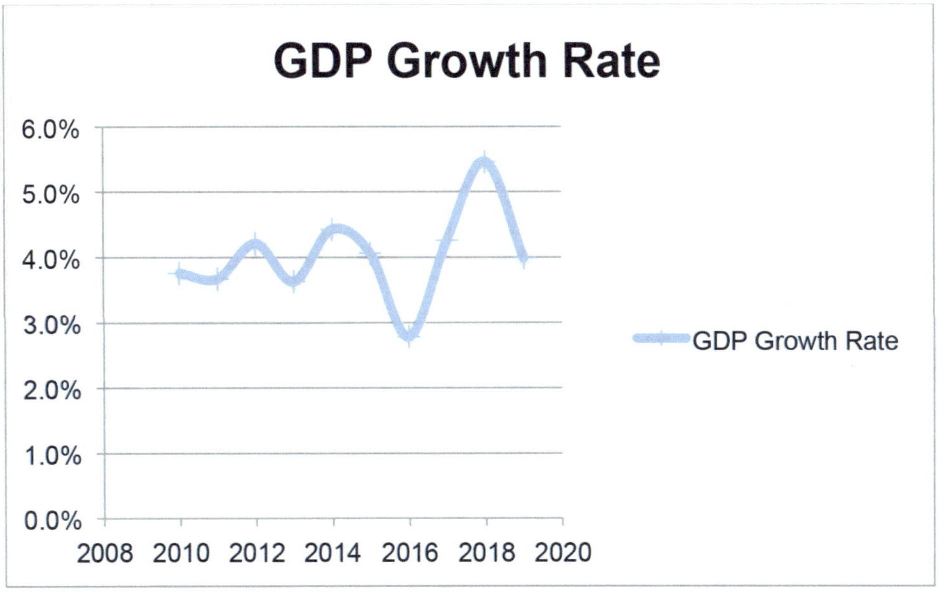

Figure 7.2.4.1 GDP Growth Rate for the period 1990 - 2019

7.2.5 GDP per Capita

Using the USA population as noted in subsection 7.4 and the GDP value in Table 7.2.1, the GDP per Capita in 2019 is estimated approximately as USD 65,000.

Additional important economic indices are included in subsections 7.5 - 7.7.

7.3 Energy Data

Source: IEA Energy Efficiency Indicators https://www.iea.org/subscribe-to-data-services/energy-efficiency-statistics

7.3.1 World Energy Supply Statistics (1990 – 2018)

The values shown in Table 7.3.1 are derived from the source noted in the table.

Table 7.3.1 World Energy Consumption (1990-2018)

Source: IEA World Energy Balances 2020 https://www.iea.org/subscribe-to-data-services/world-energy-balances-and-statistics

Year	% of year Total Coal	% of year Total Natural gas	% of year Total Nuclear	% of year Total Hydro	% of year Total Wind, solar, etc.	% of year Total Biofuels and waste	% of year Total Oil	Total ktoe
1990	25%	19%	6%	2%	0%	10%	37%	8,766,303
1995	24%	20%	7%	2%	0%	10%	37%	9,218,314
2000	23%	21%	7%	2%	1%	10%	37%	10,032,895
2005	26%	21%	6%	2%	1%	9%	35%	11,493,833
2010	28%	21%	6%	2%	1%	9%	32%	12,843,784
2015	28%	22%	5%	2%	2%	9%	32%	13,579,849
2018	27%	23%	5%	3%	2%	9%	31%	14,279,569

Figure 7.3.1 shows plot of the total energy consumption (valuses corresponding to the last column in Table 7.3.1).

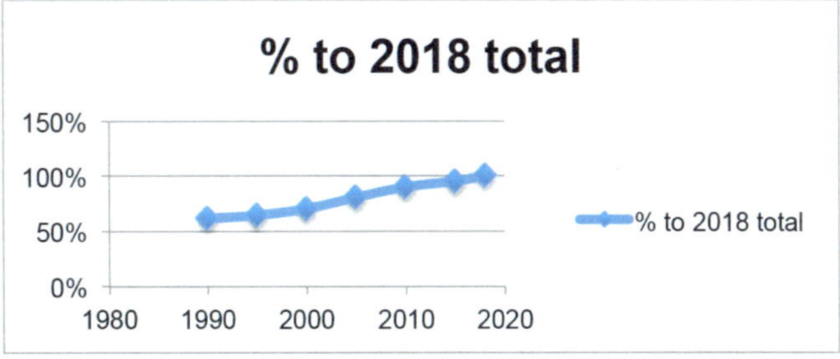

2018	Total in Ktoe	14,279,569

Figure 7.3.1.1 Plot of the World total energy consumption (valuses corresponding to the last column in Table 7.3.1.1) 1990 -2018

One notes the significant increase over the last three decades from the figure above, particularly in the last two decades.

7.3.2 Energy consumption in the USA (1990 to 2018)

Table 7.3.2 USA Energy Consumption (use) 1990 -2018

Year	% of year total Industry	% of year total T ransport	% of year total Residential	% of year total Commercial and public services	% of year total Agriculture / forestry	% of year total Non-specified	% of year total Non-energy use	% total	Total (ktoe)
1990	21.9%	37.7%	16.2%	12.3%	1.1%	1.5%	9.2%	100%	1,293,556
1995	19.8%	38.3%	18.0%	12.6%	1.1%	0.0%	10.2%	100%	1,377,747
2000	21.5%	38.0%	17.1%	12.5%	0.9%	0.1%	9.9%	100%	1,546,291
2005	17.7%	39.9%	17.4%	12.8%	1.4%	0.8%	10.0%	100%	1,563,078
2010	17.9%	39.4%	17.9%	13.7%	1.4%	0.8%	8.9%	100%	1,512,988
2015	17.4%	40.6%	16.9%	13.8%	1.4%	0.9%	8.9%	100%	1,510,648
2018	17.4%	40.0%	17.0%	13.6%	1.3%	1.3%	9.4%	100%	1,594,129

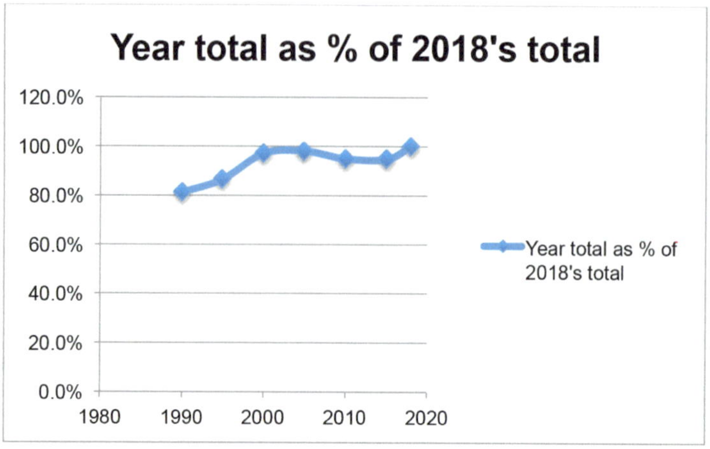

2018	Total Consumption in ktoe	1,594,129

Figure 7.3.2.1 Plot of the Total Energy Consumption of the USA 1990 -2018

From the Figure above one sees there was some decrease in the total consumption due to the energy efficiency program in the latter years.

One notes that the total consumption of the USA (this subsection) is approximately 10% of the World's (subsection 7.4.1).

7.3.3 USA Electricity Consumption (1990 to 2018)

Table 7.3.3 USA Electricity Consumption 1990 -2019

Year	% of year Total Coal	% of year Total Oil	% of year Total Natural gas	% of year Total Biofuels	% of year Total Waste	% of year Total Nuclear	% of year Total Hydro	% of year Total Geoth'l	% of year Total Solar PV	% of year Solar thermal	% of year Total Wind	% of year Total Other sources	Total GWh
1990	53%	4%	12%	2%	0%	19%	9%	0%	0%	0%	0%	0%	3,218,621
1995	51%	2%	15%	1%	1%	20%	9%	0%	0%	0%	0%	0%	3,582,114
2000	53%	3%	16%	1%	1%	20%	7%	0%	0%	0%	0%	0%	4,052,667
2005	50%	3%	18%	1%	1%	19%	7%	0%	0%	0%	0%	0%	4,294,368
2010	46%	1%	23%	1%	0%	19%	7%	0%	0%	0%	2%	0%	4,378,430
2015	34%	1%	32%	1%	0%	19%	6%	0%	1%	0%	4%	0%	4,317,159
2019	24%	1%	37%	1%	0%	19%	7%	0%	2%	0%	7%	0%	4,366,467

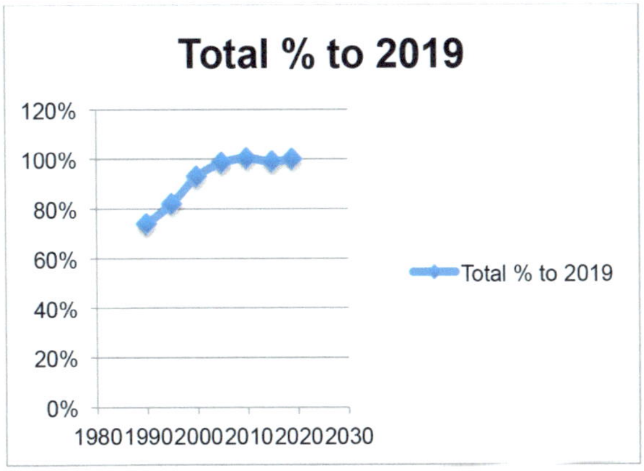

2019	Total GWh	4,366,467

Figure 7.3.3.1 Plot of the total USA electricity Consumption 1990 -2019

Please note the trend of the total consumption of the electricity between; 2017 to 2019 is like the GDP Figure 7.2.3.1

7.3.4 Electricity Consumption per Capita (1990 – 2019)

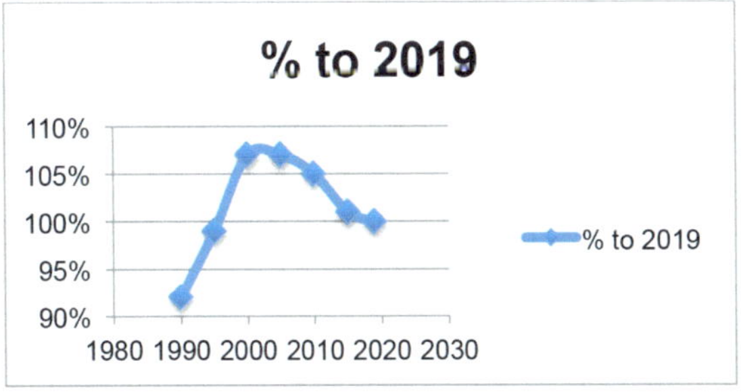

2019	12.75	MWh/capita

Figure 7.3.4.1 Plot of the annual electricity consumption per capita

The reduction in the annual electricity consumption per capita in the latter years as the energy efficiency measures appear to produce good results.

It is of interest to note per capita in 2019 electricity consumption was 12.75 MWh/ capita per year, versus the GDP/capita (from subsection 7.2. 5) is 65,000 $/year. Noting the average price of electricity to the end-consumer – Figure 7.3.4.2 below 0.137 $/kWh, the annual electricity cost to the end customer amounts to approximately 1900 $/year equivalent to about 2.8% of the GDP per Capita.

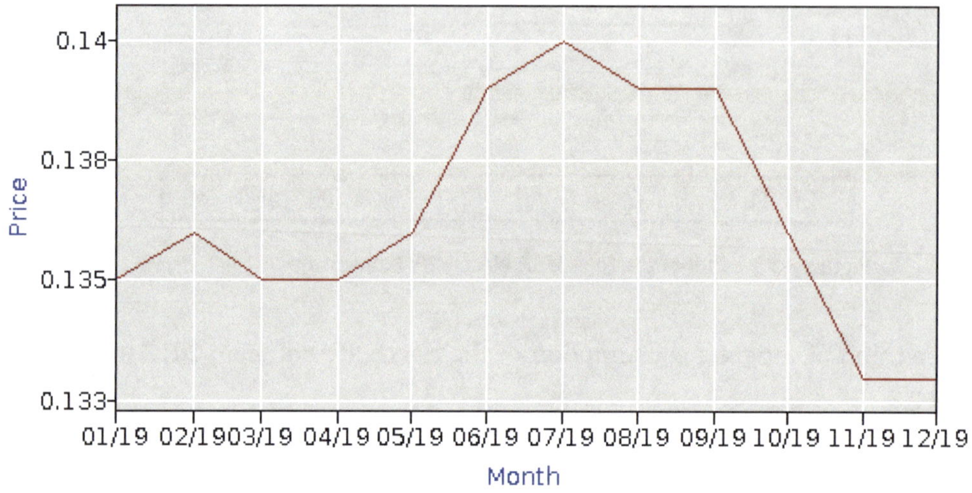

Figure 7.3.4.2 Electricity Price for 2019-2020

APU000072610

Electricity per KWH in U.S. city average, average price, not seasonally adjusted U.S. city average Electricity per KWH

<u>**U.S. BUREAU OF LABOR STATISTICS**</u>

7.3.5 USA Electricity in energy intensive manufacturing.

Table 7.3.5 USE of Electricity in Energy Intensive Industries (2000 -2017)

Year	Manufacturing energy intensity per value added (MJ/USD PPP 2015)	Manufacturing energy intensity per value added (kWh/USD PPP 2015)	% to 2017
2000	8.79	2.44	174%
2005	6.78	1.88	134%
2010	5.65	1.57	112%
2015	5.24	1.46	104%
2017	5.05	1.40	100%

The values / trend shown in the third column and the last column are quite interesting showing the energy efficiency program is giving good results.

7.3.6 USA Energy Consumption by end use

Table 7.3.6.1 Energy End-use Consumption (2000 -2018) Values are in PJ

Year	Residential Space Heating	Residential Space Cooling	Residential Appliances	Sum
2000	6,083.78	514.06	2,014.35	8,612.19
2005	5,300.24	758.36	2,466.10	8,524.70
2010	5,487.37	788.77	2,550.23	8,826.37
2015	5,185.54	762.99	2,620.98	8,569.51
2018	5,372.32	846.74	2,704.66	8,923.72

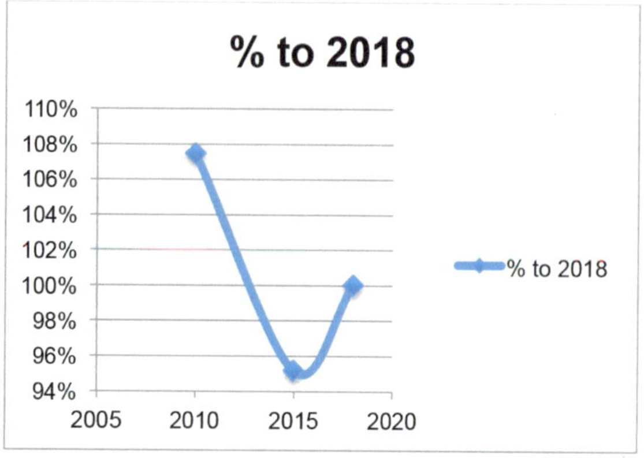

2018	Total in PJ	3,389

Figure 7.3.6.1 Energy Consumption End-use (2000 – 2018)

Table 7.3.6.2 Types of End-use Consumption (2000 – 2018)

Year	% to 2018 Residential Space Heating	% to 2018 Residential Space Cooling	% to 2018 Residential Appliances
2000	113%	61%	74%
2005	99%	90%	91%
2010	102%	93%	94%
2015	97%	90%	97%
2018	100%	100%	100%

2018 in PJ	5,372.32	846.74	2,704.66
% to Sum 2018	60.2%	9.5%	30.3%

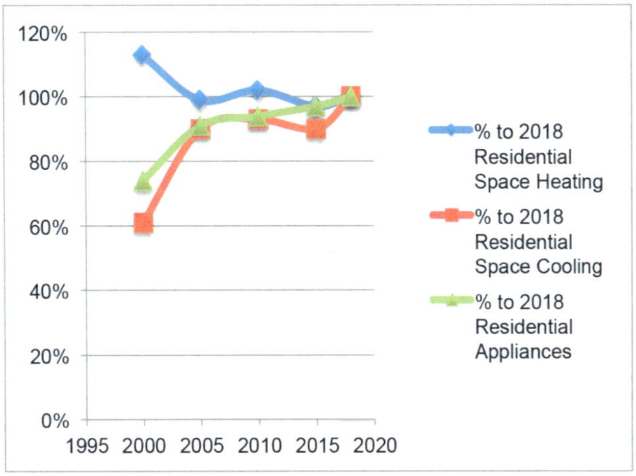

Figure 7.3.6.2 Plots of Type of use End-Use energy consumption

7.3.7 USA Transport Energy Consumption

Table 7.3.7 Energy Consumption Transport (2000 – 2018)

Values are in PJ

Year	Cars/Light Trucks	Passenger transport	Trucks	Freight transport	Sum
2000	15,675	19,026	4,480	5,281	44,462
2005	16,860	19,958	5,173	6,018	48,009
2010	15,644	18,238	5,000	5,739	44,621
2015	16,668	19,434	5,519	6,295	47,916
2018	16,279	19,282	5,736	6,482	47,779

Year	% of year total Cars/ Light Trucks	% of year total Passenger transport	% of year total Trucks	% of year total Freight transport	Sum
2000	35.3%	42.8%	10.1%	11.9%	100.0%
2005	35.1%	41.6%	10.8%	12.5%	100.0%
2010	35.1%	40.9%	11.2%	12.9%	100.0%
2015	34.8%	40.6%	11.5%	13.1%	100.0%
2018	34.1%	40.4%	12.0%	13.6%	100.0%

7.3.8 USA Gasoline Price

Table 7.3.8 Price of Gasoline 1994 -2019

Year	Mid-grade motor gasoline (USD/liter)
1994	0.31
1995	0.32
2000	0.42
2005	0.63
2010	0.77
2015	0.70
2019	0.79

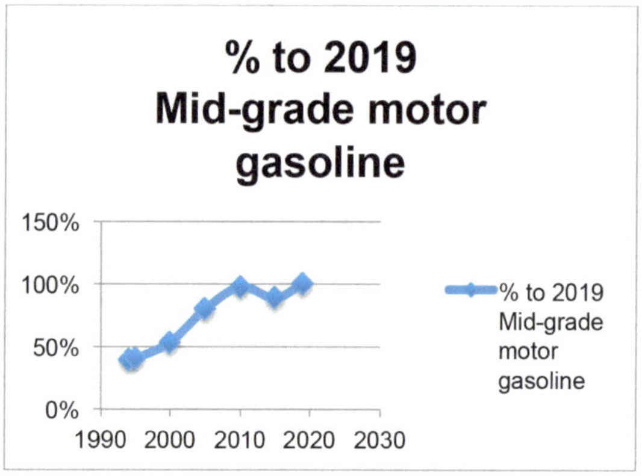

2019	USD/liter	0.79

Figure 7.3.8 Price of Gasoline 1994-2019

7.3.9 USA Natural Gas Consumption (1990 -2018)

The use of Natural Gas has been steadily increasing in recent years as compared to coal and oil it has lower CO_2 emission.

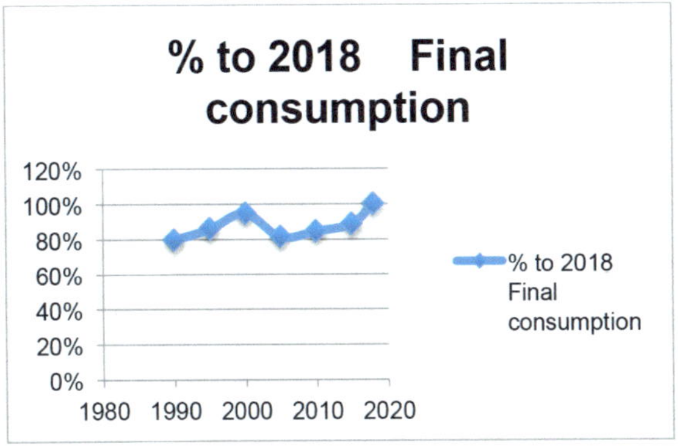

2018	Final Consumption TJ-Gross	7,705,298

Figure 7.3.9 Final Natural Gas Consumption 1990 -2018

Source:IEAGasInformation2020https://www.iea.org/subscribe-to-data-services/natural-gas-statistics

7.4 Labor Data

Table 7.4.0 Employment by major occupational group, 2019 and projected 2029 (Numbers in thousands)

2019 National Employment Matrix title and code		Employment		Change, 2019-29		Median annual wage, 2019(1)
		2019	2029	Number	Percent	
Total, all occupations	00-0000	162,795.60	168,834.70	6,039.20	3.7	$39,810
Management occupations	11-0000	10,697.20	11,202.20	505	4.7	$105,660
Business and financial operations occupations	13-0000	9,016.30	9,492.50	476.2	5.3	$69,820
Computer and mathematical occupations	15-0000	4,845.10	5,432.40	587.4	12.1	$88,340
Architecture and engineering occupations	17-0000	2,728.90	2,803.70	74.8	2.7	$81,440

Life, physical, and social science occupations	19-0000	1,460.30	1,528.50	68.2	4.7	$68,160
Community and social service occupations	21-0000	2,797.70	3,146.30	348.6	12.5	$46,090
Legal occupations	23-0000	1,335.60	1,403.20	67.6	5.1	$81,820
Educational instruction and library occupations	25-0000	9,744.70	10,185.60	441	4.5	$50,790
Arts, design, entertainment, sports, and media occupations	27-0000	2,923.60	2,997.60	73.9	2.5	$51,150
Healthcare practitioners and technical occupations	29-0000	9,133.70	9,967.30	833.6	9.1	$68,190
Healthcare support occupations	31-0000	7,013.40	8,598.10	1,584.80	22.6	$28,470
Protective service occupations	33-0000	3,615.90	3,711.10	95.2	2.6	$41,580
Food preparation and serving related occupations	35-0000	13,761.30	14,769.70	1,008.40	7.3	$24,220
Building and grounds cleaning and maintenance occupations	37-0000	5,664.00	5,943.40	279.5	4.9	$28,330
Personal care and service occupations	39-0000	4,724.10	5,088.40	364.3	7.7	$26,220
Sales and related occupations	41-0000	15,525.30	15,222.00	-303.4	-2	$29,630
Office and administrative support occupations	43-0000	20,633.00	19,673.30	-959.7	-4.7	$37,580
Farming, fishing, and forestry occupations	45-0000	1,116.60	1,115.10	-1.5	-0.1	$27,180
Construction and extraction occupations	47-0000	7,349.10	7,645.40	296.3	4	$47,430
Installation, maintenance, and repair occupations	49-0000	6,128.00	6,301.30	173.3	2.8	$46,630
Production occupations	51-0000	9,447.00	9,023.80	-423.2	-4.5	$36,000
Transportation and material moving occupations	53-0000	13,134.80	13,583.60	448.9	3.4	$32,440

7.4.1 Ratio of work Force to Population

According to the CIA USA Country report – the estimated population of the USA is 332,639,102 (July 2020).

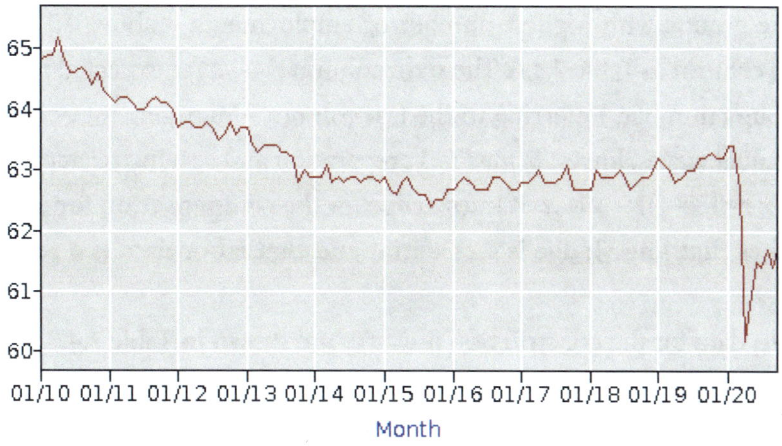

Figure 7.4.1 Plot of the Ratio of the Workforce to the Population (2010 -2020) – BLS

7.4.2 Work Force (2010 -2019)

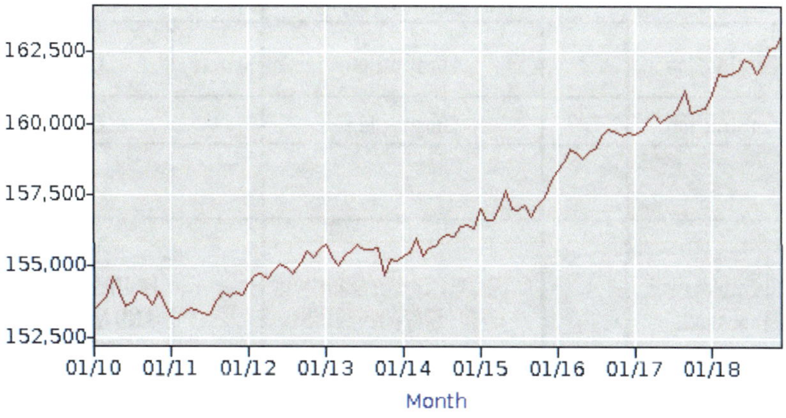

Figure 7.4.2 Labor Force – Number of the work force 2010 – 2019 - BLS

7.4.3 Employment by Major Occupational Groups

Please refer to table 7.4.0 for the Number of persons in the workforce by groups

From Table 7.4.0, it is of interest to note the average annual wage of the work force as approximately 40,000 $/year for 2019.

One notes the groups with highest number of employments – above 10 million (marked in Grey) in the third column in Table 7.4.0. The sixth column shows the expected positive and negative change in the group's number. Referring to the last column –Management occupations command the highest the annual wages, lowest is for food preparation and serving related occupations get the lowest wages. **This reflects the effect of knowledge on the compensation for employees.**

This underlines that knowledge is a essential and that labor data is a reasonable indicator for the knowledge.

As an example, data for the construction industry are shown in Table 7.4.3.1

Table 7.4.3 Sample: Construction group – data 2010 to 2020

Series Id:	**CES2000000001**		**CES2000000003**		**CES2000000007**	
Seasonally Adjusted						
Series Title:	All employees, thousands, construction, seasonally adjusted		Average hourly earnings of all employees, construction, sea Average hourly earnings of all employees, construction, seasonally adjusted		Average weekly hours of production and nonsupervisory employees, construction, seasonally adjusted	
Super Sector:	Construction		Construction		Construction	
Industry:	Construction		Construction		Construction	
NAICS Code:	23		23		23	
Data Type:	All Employees, Thousands		AVERAGE HOURLY EARNINGS OF ALL EMPLOYEES $/h		AVERAGE WEEKLY HOURS OF PRODUCTION AND NONSUPERVISORY EMPLOYEES $/h	
Years:	**2010 to 2020**					
Year	**Jan**	**Jul**	**Jan**	**Jul**	**Jan**	**Jul**
2010	5580	5508	25.12	25.17	37.8	38.2
2011	5427	5547	25.46	25.36	37.7	39.0

2012	5626	5632	25.50	25.75	39.2	39.1
2013	5746	5859	25.96	26.17	39.6	39.4
2014	5985	6178	26.41	26.66	39.0	39.8
2015	6320	6472	27.07	27.36	39.7	39.6
2016	6620	6736	27.66	28.20	39.9	39.7
2017	6847	6960	28.53	28.95	39.4	39.6
2018	7114	7304	29.39	29.94	39.4	40.2
2019	7452	7504	30.32	30.75	40.5	39.6
2020	7593	7202	31.19	31.74	40.0	39.3

7.4.4 Occupations with the most Job Growth

Table 7.4.4 Occupations with the Most Job Growth

Occupations with the most job growth, 2019 and projected 2029 (Numbers in thousands)

2019 National Employment Matrix title and code		**Employment**		**Change, 2019-29**		**Median annual wage, 2019(1)**
		2019	**2029**	**Number**	**Percent**	
Total, all occupations	00-0000	162,795.60	168,834.70	6,039.20	3.7	$39,810
Home health and personal care aides	31-1120	3,439.70	4,599.20	1,159.50	33.7	$25,280
Fast food and counter workers	35-3023	4,047.70	4,508.60	460.9	11.4	$22,740
Cooks, restaurant	35-2014	1,417.30	1,744.60	327.3	23.1	$27,790
Software developers and software quality assurance analysts and testers	15-1256	1,469.20	1,785.20	316	21.5	$107,510
Registered nurses	29-1141	3,096.70	3,318.70	221.9	7.2	$73,300
General and operations managers	11-1021	2,486.40	2,630.20	143.8	5.8	$100,780
Medical assistants	31-9092	725.2	864.4	139.2	19.2	$34,800
Medical and health services managers	Nov-11	422.3	555.5	133.2	31.5	$100,980
Market research analysts and marketing specialists	13-1161	738.1	868.4	130.3	17.7	$63,790

Laborers and freight, stock, and material movers, hand	53-7062	2,986.00	3,111.70	125.7	4.2	$29,510
Landscaping and grounds keeping workers	37-3011	1,188.00	1,307.90	119.9	10.1	$30,440
Nursing assistants	31-1131	1,528.50	1,645.50	116.9	7.6	$29,660
Nurse practitioners	29-1171	211.3	322	110.7	52.4	$109,820
Financial managers	Nov-31	697.9	806	108.1	15.5	$129,890
Janitors and cleaners, except maids and housekeeping cleaners	37-2011	2,374.20	2,479.80	105.6	4.4	$27,430
Waiters and waitresses	35-3031	2,613.80	2,711.40	97.6	3.7	$22,890
Passenger vehicle drivers, except bus drivers, transit and intercity	53-3058	853.3	947.8	94.4	11.1	$31,340
Management analysts	13-1111	876.3	970.2	93.8	10.7	$85,260
Project management specialists and business operations specialists, all other	13-1198	1,361.80	1,441.60	79.8	5.9	$73,570
Substance abuse, behavioral disorder, and mental health counselors	21-1018	319.4	398.4	79	24.7	$46,240
Construction laborers	47-2061	1,398.00	1,473.40	75.4	5.4	$36,860
Social and human service assistants	21-1093	425.6	497.1	71.5	16.8	$35,060
Animal caretakers	39-2021	300.7	369.5	68.8	22.9	$24,780
Licensed practical and licensed vocational nurses	29-2061	721.7	787.4	65.7	9.1	$47,480
Sales representatives of services, except advertising, insurance, financial services, and travel	41-3091	1,070.50	1,134.70	64.2	6	$56,130
Maintenance and repair workers, general	49-9071	1,516.40	1,579.40	63	4.2	$39,080
Industrial machinery mechanics	49-9041	399.4	461.7	62.3	15.6	$53,590
Electricians	47-2111	739.2	801.4	62.2	8.4	$56,180
Accountants and auditors	13-2011	1,436.10	1,497.90	61.7	4.3	$71,550
First-line supervisors of food preparation and serving workers	35-1012	1,039.30	1,100.20	60.9	5.9	$33,400

In the column of the forecasted % growth, the occupations with double digit growth over ten years 2019-2029 have been underlined. One notes many of these reflect society evolution.

7.4.5 Fastest Growing Occupations

Table 7.4.5 Fastest Growing Occupations

Fastest growing occupations, 2019 and projected 2029 (Numbers in thousands)

2019 National Employment Matrix title and code		Employment		Change, 2019-29		Median annual wage, 2019(1)
		2019	2029	Number	Percent	
Total, all occupations	00-0000	162,795.60	168,834.70	6,039.20	3.7	$39,810
Wind turbine service technicians	49-9081	7	11.3	4.3	60.7	$52,910
Nurse practitioners	29-1171	211.3	322	110.7	52.4	$109,820
Solar photovoltaic installers	47-2231	12	18.1	6.1	50.5	$44,890
Occupational therapy assistants	31-2011	47.1	63.5	16.3	34.6	$61,510
Statisticians	15-2041	42.7	57.5	14.8	34.6	$91,160
Home health and personal care aides	31-1120	3,439.70	4,599.20	1,159.50	33.7	$25,280
Physical therapist assistants	31-2021	98.7	130.9	32.2	32.6	$58,790
Medical and health services managers	Nov-11	422.3	555.5	133.2	31.5	$100,980
Physician assistants	29-1071	125.5	164.8	39.3	31.3	$112,260
Information security analysts	15-1212	131	171.9	40.9	31.2	$99,730
Data scientists and mathematical science occupations, all other	15-2098	33.2	43.4	10.3	30.9	$94,280
Derrick operators, oil and gas	47-5011	12	15.7	3.7	30.5	$46,990

Rotary drill operators, oil and gas	47-5012	20.9	26.6	5.6	26.9	$54,980
Roustabouts, oil and gas	47-5071	58.5	73.1	14.7	25.1	$38,910
Speech-language pathologists	29-1127	162.6	203.1	40.5	24.9	$79,120
Operations research analysts	15-2031	105.1	131.3	26.1	24.8	$84,810
Substance abuse, behavioral disorder, and mental health counselors	21-1018	319.4	398.4	79	24.7	$46,240
Forest fire inspectors and prevention specialists	33-2022	2.3	2.8	0.5	24.3	$45,270
Cooks, restaurant	35-2014	1,417.30	1,744.60	327.3	23.1	$27,790
Animal caretakers	39-2021	300.7	369.5	68.8	22.9	$24,780
Service unit operators, oil and gas	47-5013	51.7	63.6	11.8	22.9	$46,740
Marriage and family therapists	21-1013	66.2	80.9	14.8	22.3	$49,610
Computer numerically controlled tool programmers	51-9162	25.7	31.3	5.6	21.9	$56,450
Film and video editors	27-4032	38.3	46.5	8.3	21.6	$63,780
Software developers and software quality assurance analysts and testers	15-1256	1,469.20	1,785.20	316	21.5	$107,510
Genetic counselors	29-9092	2.6	3.2	0.6	21.5	$81,880
Physical therapist aides	31-2022	50.6	61.3	10.8	21.3	$27,000
Massage therapists	31-9011	166.7	201.1	34.4	20.6	$42,820
Health specialties teachers, postsecondary	25-1071	254	306.1	52.1	20.5	$97,320
Helpers--extraction workers	47-5081	16.9	20.3	3.4	20.2	$37,120

It is interesting to note the occupations with the highest values for the fastest growth, in the column percent. Knowledge is in the center of occupations with fastest change rate.

Four industries were selected for the discussions in the following subsections (7.4.6 through 7.4.13): Manufacturing, Construction, Financial Services, and Information Services. The discussions are to shed light on the relation of labor and knowledge. Please note the industries that were addressed in section 6 – "application of the K&E model".

7.4.6 Workforce Statistics (4 selected industries)

In Table 7.4.6, one can see the following parameters for the four industries:

- o Employment
- o Employment excluding the supervision
- o Unemployment
- o Job openings
- o Hires
- o Separations

Table 7.4.6 Workforce Statistics – Monthly 2020

Workforce Statistics

Employment, Unemployment, and Openings, Hires, and Separations				
Data series	Jun. 2020	Jul. 2020	Aug. 2020	Sep. 2020
Employment (in thousands)				
Employment, all employees (seasonally adjusted)				
Mnfg	12,062	12,103	(P)12,139	(P)12,205
Const'n	7,171	7,202	(P)7,219	(P)7,245
FIN	8,605	8,620	(P)8,646	(P)8,683
INF.	2,576	2,565	(P)2,591	(P)2,618

Employment, production and nonsupervisory employees (seasonally adjusted)				
Mnfg	8,370	8,408	(P)8,404	(P)8,438
Const'n	5,292	5,297	(P)5,312	(P)5,329
FIN	6,571	6,567	(P)6,575	(P)6,592
INF.	2,032	2,041	(P)2,049	(P)2,067
Unemployment				
Unemployment rate				
Mnfg	9.10%	8.60%	6.70%	6.20%
Const'n	10.10%	8.90%	7.60%	7.10%
FIN	5.10%	4.70%	4.20%	4.40%
INF.	12.00%	12.30%	8.60%	8.60%
Job openings, hires, and separations (in thousands)				
Job openings				
Mnfg	346	430	(P)460	
Const'n	244	332	(P)264	
FIN	293	(P)276		
INF.	97	106	(P)81	-
Hires				
Mnfg	503	379	(P)428	
Const'n	581	432	(P)388	
FIN	222	267	(P)215	
INF.	61	74	(P)77	-
Separations				
Mnfg	402	384	(P)408	
Const'n	317	367	(P)354	
FIN	183	204	(P)251	
INF.	52	60	(P)69	-

The data in Table 7.4.6.1 reflects the level of activities as related to the labor-force in the four industries. P stands for preliminary.

7.4.7 Dynamic Changes in Employment (4 selected industries)

Table 7.4.7 Dynamic Changes in employment 2019-2020

Data series	2nd quarter 2019	3rd quarter 2019	4th quarter 2019	1st quarter 2020
Dynamic changes in employment (in thousands)				
Gross job gains				
Mnfg	416	377	409	364
Mnfg	428	441	407	418
Const'n	690	649	676	683
Const'n	673	647	649	626
FIN	410	386	406	356
FIN	364	364	362	363
INF	164	162	155	152
INF	163	154	161	139

Table 7.4.7 above shows the seasonal variation.

7.4.8 Employment by Occupation (4 selected industries)

Table 7.4.8 Employment by Occupation 2019

W	Data series	Employment, 2019
Mnfg	Helpers--production workers	186,280
Mnfg	Inspectors, testers, sorters, samplers, and weighers	365,290
Mnfg	Machinists	316,630
Mnfg	Purchasing agents, except wholesale, retail, and farm products	100,630
Mnfg	Team assemblers	1,006,910
Const'n	Carpenters	621,980

Const'n	Construction laborers	866,650
Const'n	Construction managers	241,260
Const'n	Electricians	527,300
Const'n	Operating engineers and other construction equipment operators	260,380
FIN	N/A	
INF.	Customer service representatives	142,560
INF.	Telecommunications equipment installers and repairers, except line installers	145,200

The table above gives interesting view for the labor force composition in each of the selected four industries.

7.4.9 Earnings (4 selected industries)

In this subsection, the following parameters are covered:

- Average hourly earnings per hour
- Working hours per week

The top part of the Table is for all employees and the bottom is excluding the supervision.

Table 7.4.9.1 Earnings - All employees 2020

Earnings and Hours

Earnings and Hours of All Employees

	Data series	Jun. 2020	Jul. 2020	Aug. 2020	Sep. 2020
Mnfg	Average hourly earnings	$28.62	$28.77	(P)$28.93	(P)$28.85
Mnfg	Average weekly hours	39.1	39.8	(P)40.0	(P)40.2
Const'n	Average hourly earnings	$31.65	$31.74	(P)$31.83	(P)$31.83
Const'n	Average weekly hours	38.8	38.8	(P)38.8	(P)38.8
FIN	Average hourly earnings	$37.83	$37.73	(P)$38.05	(P)$38.17

FIN	Average weekly hours	37.5	37.5	(P)37.5	(P)37.6
INF.	Average hourly earnings	$43.35	$44.10	(P)$43.94	(P)$43.77
INF.	Average weekly hours	36.5	36.3	(P)36.5	(P)36.4

Earnings and Hours of Production and Nonsupervisory Employees

	Data series	Jun. 2020	Jul. 2020	Aug. 2020	Sep. 2020
Mnfg	Average hourly earnings	$22.61	$22.81	(P)$22.81	(P)$22.94
Mnfg	Average weekly hours	39.9	40.7	(P)41.0	(P)41.0
Const'n	Average hourly earnings	$29.32	$29.30	(P)$29.40	(P)$29.07
Const'n	Average weekly hours	39.4	39.3	(P)39.3	(P)39.3
FIN	Average hourly earnings	$29.03	$29.15	(P)$29.35	(P)$29.51
FIN	Average weekly hours	36.9	37.1	(P)36.8	(P)37.2
INF.	Average hourly earnings	$35.74	$35.87	(P)$35.88	(P)$35.69
INF.	Average weekly hours	36.2	35.8	(P)36.8	(P)36.2

The tables above show the average hourly earnings in $ for the all the work force in the industry and for workers excluding supervisors. The knowledge influence on the earnings can be detected in two respects, the difference when the supervision is included – different values from top part to bottom. In the second respect, how the Information occupants command relatively higher earnings (knowledge content).

7.4.10 Price Indices (4 selected industries)

In table 7.4.10 the following parameters are included:

- Producer Production Index "PPI"[1]
- Import Index
- Export Index

1 Please refer to subsection 7.7.2 for definition

Table 7.4.10 Price Indices – Change from previous month

	Data series	Jun. 2020	Jul. 2020	Aug. 2020	Sep. 2020
	Producer price index	(P)190.4	(P)192.7	(P)192.8	(P)192.9
Mnfg	**Percent change from previous month**	(P)0.8%	(P)1.2%	(P)0.1%	(P)0.1%
Mnfg	Import index	109.7	(R)110.1	111	111.4
Mnfg	**Percent change from previous month**	0.40%	(R)0.4%	(R)0.8%	0.40%
Mnfg	Export index	113.4	114.1	(R)114.7	115.1
Mnfg	**Percent change from previous month**	0.80%	0.60%	(R)0.5%	(R)0.3%
INF.	Producer price index	(P)103.7	(P)103.8	(P)104.3	(P)104.1
INF.	**Percent change from previous month**	(P)-0.1%	(P)0.1%	(P)0.5%	(P)-0.2%

Unfortunately the data was nor complete Table 7.4.10 does not have all the applicable entries for the four industries.

7.4.11 Workplace Trends (4 selected industries)

The following parameters are covered:

- o Number of Establishments
  - § Private Industry
  - § Local Government
  - § State Government
  - § Federal Government
- o Establishments with changes in Employment
  - § With job gains
  - § With job losses
  - § Private Industry
  - § Local Governement
  - § State Government
  - § Federal Government

Table 7.4.11 Workplace Trends – numbers of Establishments

Establishments

	ES	2nd quarter 2019	3rd quarter 2019	4th quarter 2019	1st quarter 2020
	Number of establishments				
Mnfg	Private industry	354,485	355,956	357,017	(P)358,664
Mnfg	Local government	41	41	42	(P)40
Mnfg	State government	23	23	23	(P)24
Mnfg	Federal government	47	47	47	(P)46
Mnfg	**Establishments with changes in employment (in thousands)**				
Mnfg	With job gains (seasonally adjusted)	94	91	93	89
Mnfg	With job losses (seasonally adjusted)	98	99	96	101
FIN	Private industry	909,748	916,789	926,344	(P)933,220
FIN	Local government	1,953	1,955	1,959	(P)1,944
FIN	State government	114	114	115	(P)114
FIN	Federal government	785	790	783	(P)775
FIN	**Establishments with changes in employment (in thousands)**				
FIN	With job gains (seasonally adjusted)	157	157	164	150
FIN	With job losses (seasonally adjusted)	155	152	152	159
INF.	**Number of establishments**				
INF.	Private industry	183,302	186,163	189,801	(P)193,016
INF.	Local government	5,111	5,112	5,117	(P)5,115
INF.	State government	320	323	321	(P)320
INF.	Federal government	126	127	126	(P)126
INF.	**Establishments with changes in employment (in thousands)**				
INF.	With job gains (seasonally adjusted)	36	36	37	36
INF.	With job losses (seasonally adjusted)	37	37	37	38

Data extracted on: October 30, 2020

The changes in the numbers of the establishments in the various industries reflect the dynamics of the industry in year 2019/2020. Data for construction were not available.

7.4.12 Productivity (one selected industry)

Table 7.4.12 Productivity in "Output Indices" Manufacturing - year 2019/2020

Productivity

	Data series	3rd quarter 2019	4th quarter 2019	1st quarter 2020	2nd quarter 2020
Mnfg	Output per hour index (seasonally adjusted)	98.517	98.34	(R)98.759	(R)94.950
Mnfg	**Percent change from previous quarter**	-0.5	-0.7	(R)1.7	(R)-14.6
Mnfg	Hours index (seasonally adjusted)	106.469	106.604	104.684	(R)92.896
Mnfg	**Percent change from previous quarter**	1.4	0.5	-7	(R)-38.0
Mnfg	Output index (seasonally adjusted)	104.89	104.835	(R)103.385	(R)88.205
Mnfg	**Percent change from previous quarter**	0.9	-0.2	(R)-5.4	-47
Mnfg	Unit labor costs (seasonally adjusted)	116.694	118.399	(R)119.618	(R)127.510
Mnfg	**Percent change from previous quarter**	-1.4	6	(R)4.2	(R)29.1

In subsection 7.4 salient labor data have been presented. It is very important to recognize that knowledge could be correlated to many aspects of the activities of the labor. As was pointed out in section 5, the origin of knowledge is Humans. The word labor may not be a good word that reflects the value added due to the contributions of the labor-force.

7.5 Fiscal and Monetary Policy

7.5.1 Federal Reserve Board – Debt / GDP Historical Data (2009 -2019)

From the Federal Reserve Board data, the history of the debt in relation to the GDP can be seen. The original data is on quarterly basis. For each year the average for the year was calculated as the arithmetic average of the four quarters of the year. The calculated results are included in Table 7.5.1.1

Table 7.5.1.1 Shows part of the calculated data - For debts

- Nonfinancial Business
- Households and Nonprofits
- Federal Government
- State and Local Governments

Table 7.5.1.1 Debt Summary & Average GDP (2009 – 2019)

	DEBT					
Year	Total Nonfinancial	Households and Nonprofits	Nonfinancial Business	Federal Government	State and Local Governments	Avg GDP
2019	53,542,555	15,799,708	16,009,666	18,656,876	3,076,306	21,433,226
2018	51,141,289	15,326,076	15,152,287	17,560,391	3,102,536	20,611,861
2017	48,485,091	14,796,423	14,181,411	16,390,520	3,116,737	19,542,980
2016	46,597,189	14,315,850	13,367,781	15,786,216	3,127,342	18,745,076
2015	44,565,935	14,030,584	12,588,279	14,841,203	3,105,869	18,238,301
2014	42,982,089	13,883,115	11,746,843	14,232,305	3,119,826	17,527,258
2013	41,318,215	13,648,169	11,093,055	13,414,439	3,162,552	16,784,851
2012	39,838,393	13,595,505	10,554,847	12,495,746	3,192,295	16,197,007
2011	38,241,199	13,702,282	10,133,155	11,222,531	3,183,231	15,542,582

2010	37,048,366	13,814,221	10,075,974	9,983,009	3,175,163	14,992,052
2009	35,864,092	13,998,699	10,404,262	8,390,183	3,070,949	14,448,933

The above data are shown graphically in Figure 7.5.1 below. One can see the parameters that have grown faster than others.

Further it appears in recent years, that the sum of the debts of the Federal Government and the State & Local Governments is of the same order of magnitude as the GDP. Also, debts of the nonfinancial business, and households and nonprofits are comparable.

Figure 7.5.1 Debts and GDP 20009-2019

The increased debt may cause a negative effect on the nation's economy, particularly if it goes beyond

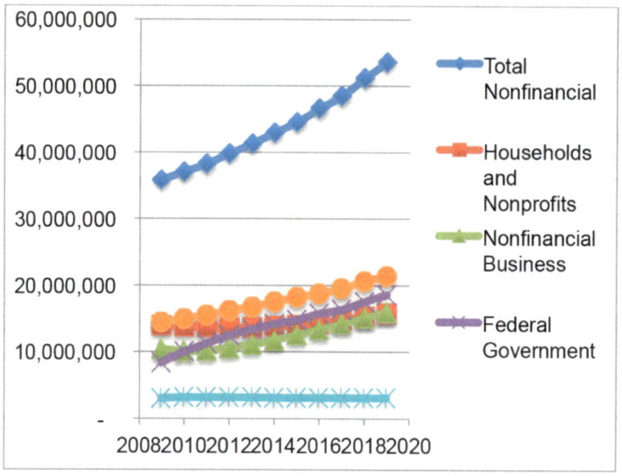

acceptable threshold. This is the reason why the Federal Reserve Board is continuously following the debts.

7.5.2 Federal Reserve Board – Assets /Income Historical Data (2009 -2019)

Again, data was extracted from the data of the Federal Reserve Board. It contains a lot of parameters. All relate to assets, net worth, and liabilities. As an example the following parameters were chosen and are included in Table 7.5.2.1

- Households and nonprofit organizations; Net Worth (million $)
- Households and nonprofit organizations; Total Assets (million $)
- Households and nonprofit organizations; Total Liabilities (million $)
- Households and nonprofit organizations; Nonfinancial Assets (million $)
- Households and nonprofit organizations; Financial Assets

Table 7.5.2.1 Assets & Disposable Personal Income (2009 – 2019)

		NET WORTH & ASSETS				
Year	Disposable Personal Income (DPI) (million $)	Households and nonprofit organizations; Net Worth (million $)	Households and nonprofit organizations; Total Assets (million $)	Households and nonprofit organizations; Total Liabilities (million $)	Households and nonprofit organizations; Nonfinancial Assets (million $)	Households and nonprofit organizations; Financial Assets
2020	17,599,731	118,210,196	134,810,850	16,600,655	40,971,573	93,839,278
2019	16,348,626	114,592,949	130,752,794	16,159,845	39,344,649	91,408,145
2018	15,766,547	107,006,846	122,685,908	15,679,062	37,114,594	85,571,314
2017	14,901,876	101,589,550	116,730,587	15,141,037	35,045,017	81,685,569
2016	14,202,828	94,233,849	108,888,919	14,655,070	32,787,771	76,101,149
2015	13,784,292	89,954,984	104,313,263	14,358,279	30,898,235	73,415,028
2014	13,207,138	85,525,360	99,715,183	14,189,823	28,984,517	70,730,666
2013	12,504,723	78,496,879	92,437,600	13,940,721	27,198,212	65,239,388
2012	12,500,626	71,079,687	84,970,956	13,891,270	25,422,977	59,547,979
2011	11,873,078	67,439,009	81,424,846	13,985,837	25,231,745	56,193,101
2010	11,313,977	64,852,494	78,949,006	14,096,511	25,775,598	53,173,408
2009	10,906,731	61,643,056	75,924,946	14,281,890	26,572,407	49,352,539
2008	10,934,709	65,550,243	80,121,302	14,571,059	28,884,517	51,236,786
2007	10,515,293	70,700,732	84,841,270	14,140,537	30,807,508	54,033,762
2006	10,024,746	67,405,544	80,510,967	13,105,422	30,620,368	49,890,599
2005	9,385,791	62,092,125	73,871,387	11,779,263	28,530,000	45,341,387
2004	8,986,333	56,063,938	66,636,660	10,572,721	25,207,604	41,429,056
2003	8,484,009	49,159,449	58,643,632	9,484,183	22,449,789	36,193,843
2002	8,106,785	46,244,105	54,715,147	8,471,042	20,622,478	34,092,669
2001	7,766,545	45,314,222	53,072,205	7,757,983	18,946,668	34,125,537
2000	7,416,272	44,623,869	51,795,481	7,171,612	16,780,791	35,014,690

1999	6,889,955	41,038,231	47,595,805	6,557,574	15,026,259	32,569,546

As the Federal Reserve Board is concerned about the debt, it needs to see the assets and liabilities. That is why it follows their pertinent parameters.

The contents of subsections 7.5.1 and 7.5.2 provide data as part of the analyses the Federal Reserve Board considers regularly for making any change of the interest rate – please refer to subsection 7.5.3

7.5.3 Interest Rate

The Federal Reserve Board reviews regularly the state of the economy and makes adjustment to the base interest-rate as the board perceives the direction of the economy. This mostly is the case of an overheating economy. To curb inflation, it increases the interest rate. On the other hand, if it sees a recession is coming due to slowing economy, it will lower the base interest rate.

Please refer to examples of economic monthly reports issued by the treasury department:

U.S. ECONOMIC STATISTICS - MONTHLY DATA

8/29/18

	2013	2014	2015	2016	2017	Feb	Mar	Apr	May	Jun	Jul
	(Annual Average)										
ISM Composite Index - Manufacturing	53.4	55.5	51.3	51.4	57.4	60.8	59.3	57.3	58.7	60.2	58.1
ISM Business Activity Index - Nonmanufacturing	56.6	59.6	60.8	58.0	60.1	62.8	60.6	59.1	61.3	63.9	56.5
	(Dec to Dec)										
New Orders for Durables (advance report, growth, percent)	0.0	-2.6	-2.0	0.9	10.5	4.5	2.7	-1.0	-0.3	0.7	-1.7
New Orders for Nondefense Capital Goods	8.7	-17.4	-12.5	5.7	18.0	3.5	8.7	-6.7	-2.3	2.0	-4.6
Business Inventories (percent change)	3.9	3.5	1.7	1.8	3.5	0.56	-0.11	0.30	0.32	0.09	
	(Annual Average)										
Business Inventories/sales ratio	1.29	1.31	1.39	1.42	1.38	1.36	1.35	1.35	1.34	1.33	
Manufacturing	1.30	1.31	1.39	1.41	1.37	1.36	1.35	1.35	1.35	1.33	
Wholesale Trade	1.19	1.21	1.33	1.35	1.30	1.29	1.29	1.27	1.25	1.25	
Retail	1.41	1.43	1.46	1.49	1.47	1.46	1.44	1.44	1.43	1.43	
U.S. Trade Balance (billions of dollars, BOP Basis)	(Annualized Average)										
Goods and services	-461.1	-489.6	-498.5	-502.0	-552.3	-55.5	-47.2	-46.1	-43.2	-46.3	
Goods	-700.5	-749.9	-761.9	-751.1	-807.5	-76.7	-69.3	-68.4	-65.7	-68.8	
	(Dec to Dec)										
Index of Leading Indicators (percent change)	4.6	6.0	2.4	1.6	5.8	0.6	0.3	0.5	0.1	0.5	0.6
Index of Coincident Indicators (percent change)	0.6	3.0	1.3	1.3	2.2	0.3	0.2	0.3	0.1	0.3	0.2
Interest Rates (percent)	(Annual Average)										
3-month T-bill	0.06	0.04	0.06	0.33	0.94	1.53	1.70	1.76	1.87	1.91	1.96
10-year T-note	2.35	2.54	2.14	1.84	2.33	2.86	2.84	2.87	2.98	2.91	2.89
10-year corporate bond yield*	3.69	3.61	3.51	3.27	3.42	3.76	3.82	3.88	4.02	4.03	4.04

*Calculated by the U.S. Treasury to provide discount rates required by the Pension Protection Act of 2006. The methodology uses a set of US corporate bonds that cover the AAA, AA and A markets

U.S. ECONOMIC STATISTICS - MONTHLY DATA

8/29/18

	2013	2014	2015	2016	2017	Feb	Mar	Apr	May	Jun	Jul
	(Annual Average)										
Unemployment Rate (level)	7.4	6.2	5.3	4.9	4.4	4.1	4.1	3.9	3.8	4.0	3.9
Payroll Employment (monthly increase, thousands)	(Annualized Average of months)										
Total Nonfarm	192	250	226	195	182	324	155	175	268	248	157
Private	197	240	213	178	180	321	153	174	260	234	170
Inflation (percent)	(Dec to Dec)										
CPI (over month)						0.2	-0.1	0.2	0.2	0.1	0.2
CPI (over year)	1.5	0.7	0.7	2.1	2.1	2.2	2.4	2.5	2.8	2.9	2.9
excluding food and energy (over month)						0.2	0.2	0.1	0.2	0.2	0.2
excluding food and energy (over year)	1.7	1.6	2.1	2.2	1.8	1.8	2.1	2.1	2.2	2.3	2.4
PCE (over year)	1.5	0.0	0.4	1.7	1.8	1.9	2.1	2.0	2.2	2.2	
PCE core (excluding food and energy, over year)	1.6	1.5	1.3	1.9	1.6	1.7	2.0	1.9	1.9	1.9	
PPI - finished goods (over month)						0.0	0.2	-0.1	1.0	0.0	0.2
PPI - finished goods (over year)	1.5	-0.8	-2.8	1.9	3.2	2.8	3.0	2.3	3.9	3.9	4.1
	(Annual Average)										
West Texas Intermediate crude oil ($/barrel, spot)	97.9	93.3	48.7	43.2	50.9	62	63	66	70	67	71
Housing (thousand units, annual rate)	(Annual Average)										
Housing Starts	928	999	1,107	1,177	1,208	1290	1327	1276	1329	1158	1168
New Single-Family Homes Sold	430	440	502	560	616	663	672	633	654	638	627
Auto and Light Truck Sales (million units, ann'l rate)	15.6	16.5	17.5	17.5	17.2	17.0	17.3	17.1	17.2	17.2	16.8
	(Dec to Dec)										
Retail Sales and Food Services (growth, percent)	3.5	3.7	2.9	3.7	5.3	0.1	0.7	0.3	1.2	0.2	0.5
ex - motor vehicles and parts dealers	3.0	2.9	1.9	2.8	6.0	0.4	0.4	0.4	1.4	0.2	0.6
Industrial Production (growth, percent)	(Dec to Dec)										
Total	2.3	3.2	-3.9	0.5	2.9	0.5	0.6	1.1	-0.8	1.0	0.1
Manufacturing	0.3	1.5	-1.9	0.4	1.7	1.5	-0.1	0.7	-1.0	0.9	0.3
Capacity Utilization (percent)	(Annual Average)										
Total	77.6	79.0	77.3	75.4	76.1	77.2	77.5	78.3	77.5	78.1	78.1
Manufacturing	75.0	75.8	75.8	74.6	74.8	75.7	75.6	76.0	75.2	75.7	75.9

The treasury department publishes the following Interest Rate Statistics. **"Interest Rate Statistics**

DAILY TREASURY YIELD CURVE RATES

This curve, which relates the yield on a security to its time to maturity is based on the closing market bid yields on actively traded Treasury securities in the over-the-counter market. These market yields are calculated from composites of quotations obtained by the Federal Reserve Bank of New York. For information on how the Treasury's yield curve is derived, visit our Treasury Yield Curve Methodology page.

DAILY TREASURY REAL YIELD CURVE RATES

The real curve, which relates the real yield on a Treasury Inflation Protected Security (TIPS) to its time to maturity, is based on the closing market real bid yields on actively traded TIPS in the over-the-counter market. These market real yields are calculated from composites of quotations obtained by the Federal Reserve Bank of New York. Treasury began publishing this series on January 2, 2004. At that time Treasury released 1 year of historical data.

DAILY TREASURY BILL RATES

These rates are composites of closing market bid quotations on recently issued Treasury Bills in the over-the-counter market as obtained by the Federal Reserve Bank of New York at approximately 3:30 PM each business day.

DAILY TREASURY LONG-TERM RATES AND EXTRAPOLATION FACTORS

Treasury ceased publication of the 30-year constant maturity series on February 18, 2002, and resumed that series on February 9, 2006. To estimate a 30-year rate during that time frame, this series includes the Treasury 20-year Constant Maturity rate and an "adjustment factor," which may be added to the 20-year rate to estimate a 30-year rate during the period in which Treasury did not sell 30-year bonds. Detailed information is provided with the data.

DAILY TREASURY REAL LONG -TERM RATE AVERAGES

Beginning on January 2, 2004, Treasury began publishing a Long-Term Real Rate Average. This series is intended for use as a proxy for long-term real rates. At that time Treasury released 1 year of historical data.

Treasury Department Publication

Selected Interest Rates (Daily) - H.15

Current Release

About

Announcements

Technical Q&As

H.15 Selected Interest Rates RSS Data Download

The release is posted daily Monday through Friday at 4:15pm. The release is not posted on holidays or in the event that the Board is closed.

Release date: December 28, 2020

Selected Interest Rates

Yields in percent per annum

Instruments	2020 Dec 21	2020 Dec 22	2020 Dec 23	2020 Dec 24	2020 Dec 25*
Federal funds (effective) 1 2 3	0.09	0.09	0.09	0.09	0.09
Commercial Paper 3 4 5 6					
Nonfinancial					
1-month	0.10	0.10	0.08	n.a.	
2-month	0.10	0.10	0.09	0.10	
3-month	0.13	0.13	0.11	0.21	
Financial					
1-month	0.08	0.12	0.12	n.a.	
2-month	n.a.	n.a.	n.a.	n.a.	
3-month	n.a.	0.21	0.17	0.14	
Bank prime loan 2 3 7	3.25	3.25	3.25	3.25	3.25
Discount window primary credit 2 8	0.25	0.25	0.25	0.25	0.25
U.S. government securities					
Treasury bills (secondary market) 3 4					
4-week	0.08	0.06	0.07	0.09	
3-month	0.09	0.09	0.09	0.09	
6-month	0.09	0.09	0.09	0.09	
1-year	0.09	0.09	0.09	0.10	
Treasury constant maturities					
Nominal 9					
1-month	0.08	0.07	0.07	0.09	
3-month	0.09	0.09	0.09	0.09	
6-month	0.09	0.09	0.09	0.09	
1-year	0.09	0.09	0.09	0.10	
2-year	0.13	0.13	0.13	0.13	
3-year	0.18	0.17	0.18	0.17	
5-year	0.39	0.37	0.38	0.37	
7-year	0.66	0.64	0.67	0.66	
10-year	0.95	0.93	0.96	0.94	
20-year	1.48	1.45	1.49	1.46	
30-year	1.68	1.65	1.70	1.66	
Inflation indexed 10					

5-year	-1.51	-1.55	-1.55	-1.55	
7-year	-1.25	-1.29	-1.28	-1.29	
10-year	-1.00	-1.01	-1.01	-1.03	
20-year	-0.56	-0.54	-0.55	-0.58	
30-year	-0.32	-0.33	-0.32	-0.35	
Inflation-indexed long-term average 11	-0.43	-0.44	-0.43	-0.46	

*	Markets closed.
n.a.	Not available.

Footnotes

1. As of March 1, 2016, the daily effective federal funds rate (EFFR) is a volume-weighted median of transaction-level data collected from depository institutions in the Report of Selected Money Market Rates (FR 2420). Prior to March 1, 2016, the EFFR was a volume-weighted mean of rates on brokered trades.

2. Weekly figures are averages of 7 calendar days ending on Wednesday of the current week; monthly figures include each calendar day in the month.

3. Annualized using a 360-day year or bank interest.

4. On a discount basis.

5. Interest rates interpolated from data on certain commercial paper trades settled by The Depository Trust Company. The trades represent sales of commercial paper by dealers or direct issuers to investors (that is, the offer side). The 1-, 2-, and 3-month rates are equivalent to the 30-, 60-, and 90-day dates reported on the Board's Commercial Paper Web page (www.federalreserve.gov/releases/cp/).

6. Financial paper that is insured by the FDIC's Temporary Liquidity Guarantee Program is not excluded from relevant indexes, nor is any financial or nonfinancial commercial paper that may be directly or indirectly affected by one or more of the Federal Reserve's liquidity facilities. Thus the rates published after September 19, 2008, likely reflect the direct or indirect effects of the new temporary programs and, accordingly, likely are not comparable for some purposes to rates published prior to that period.

7. Rate posted by a majority of top 25 (by assets in domestic offices) insured U.S.-chartered commercial banks. Prime is one of several base rates used by banks to price short-term business loans.

8. The rate charged for discounts made and advances extended under the Federal Reserve›s primary credit discount window program, which became effective January 9, 2003. This rate replaces that for adjustment credit, which was discontinued after January 8, 2003. For further information, see www.federalreserve.gov/boarddocs/press/bcreg/2002/200210312/default.htm. The rate reported is that for the Federal Reserve Bank of New York. Historical series for the rate on adjustment credit as well as the rate on primary credit are available at www.federalreserve.gov/releases/h15/data.htm.

9. Yields on actively traded non-inflation-indexed issues adjusted to constant maturities. The 30-year Treasury constant maturity series was discontinued on February 18, 2002, and reintroduced on February 9, 2006. From February 18, 2002, to February 9, 2006, the U.S. Treasury published a factor for adjusting the daily nominal 20-year constant maturity in order to estimate a 30-year nominal rate. The historical adjustment factor can be found at www.treasury.gov/resource-center/data-chart-center/interest-rates/. Source: U.S. Treasury.

11. Based on the un-weighted average bid yields for all TIPS with remaining terms to maturity of more than 10 years.

Note: Current and historical H.15 data, along with weekly, monthly, and annual averages, are available on the Board's Data Download Program (DDP) at www.federalreserve.gov/data-download/Choose.aspx?rel=H15). Weekly, monthly and annual rates are averages of business days unless otherwise noted.

Description of the Treasury Nominal and Inflation-Indexed Constant Maturity Series

Yields on Treasury nominal securities at "constant maturity" are interpolated by the U.S. Treasury from the daily yield curve for non-inflation-indexed Treasury securities. This curve, which relates the yield on a security to its time to maturity, is based on the closing market bid yields on actively traded Treasury securities in the over-the-counter market. These market yields are calculated from composites of quotations obtained by the Federal Reserve Bank of New York. The constant maturity yield values are read from the yield curve at fixed maturities, currently 1, 3, and 6 months and 1, 2, 3, 5, 7, 10, 20, and 30 years. This method provides a yield for a 10-year maturity, for example, even if no outstanding security has exactly 10 years remaining to maturity. Similarly, yields on inflation-indexed securities at "constant maturity" are interpolated from the daily yield curve for Treasury inflation protected securities in the over-the-counter market. The inflation-indexed constant maturity yields are read from this yield curve at fixed maturities, currently 5, 7, 10, 20, and 30 years.

Last Update: December 28, 2020

7.5.4 Inflation

Unlike the last two decades of the last century when the inflation rate was quite high for some time, the interest rate in the last two decades was quite tamed.

It is important to note that interest rates and inflation are quite important for making any long-term decision, mostly in investment decisions. This is discussed in section 8.

7.6 Personal Income

7.6.1 Personal Income and Its Disposition

Table 7.6.1.1 Personal Income and Its Disposition 2017 -2019

Personal Income and Its Disposition

[Billions of dollars]

Line		2017	2018	2019
1	**Personal income** [1]	**16,948.6**	**17,851.8**	**18,551.5**
2	Compensation of employees	10,422.5	10,950.1	11,432.4
3	Wages and salaries	8,471.5	8,894.2	9,309.3
4	Supplements to wages and salaries	1,951.1	2,055.9	2,123.1
5	Proprietors' income with inventory valuation and capital consumption adjustments	1,509.0	1,585.9	1,657.7
6	Farm	41.5	43.0	49.7
7	Nonfarm	1,467.4	1,542.9	1,608.0
8	Rental income of persons with capital consumption adjustment	721.9	759.3	787.1
9	Personal income receipts on assets	2,738.5	2,946.7	2,967.9
10	Personal interest income	1,577.6	1,641.6	1,677.4
11	Personal dividend income	1,160.8	1,305.1	1,290.4
12	Personal current transfer receipts	2,855.1	2,970.3	3,125.2

13	Less: Contributions for government social insurance, domestic	1,298.4	1,360.4	1,418.8
14	Less: Personal current taxes	2,046.7	2,085.3	2,202.9
15	**Equals: Disposable personal income**	**14,901.9**	**15,766.5**	**16,348.6**
16	Less: Personal outlays	13,830.9	14,529.2	15,117.4
17	**Equals: Personal saving**	**1,071.0**	**1,237.3**	**1,231.2**
18	Personal saving as a percentage of disposable personal income	7.2	7.8	7.5
	Addenda:			
19	Personal income excluding current transfer receipts, billions of chained (2012) dollars [2]	13,297.8	13,748.8	14,042.9
20	Disposable personal income, billions of chained (2012) dollars [3]	14,060.5	14,566.4	14,882.5

2. Personal income is also equal to national income less corporate profits with inventory valuation and capital consumption adjustments, taxes on production and imports less subsidies, contributions for government social insurance, net interest and miscellaneous payments, business current transfer payments (net), and current surplus of government enterprises, plus personal income receipts on assets, and personal current transfer receipts.

3. The current-dollar measure is deflated by the implicit price deflator for personal consumption expenditures.

Source: U.S. Bureau of Economic Analysis

Table 7.6.1.1 shows **Personal Income and Its Disposition** in billion USD. Please see the footnote at the bottom of the table explaining the difference of personal income (shown in Table 7.6.1.1) from national income found in Table 7.2.1.

Row # 16 in Table 7.2.1 is of interest. “Personal Outlays” as it reflects the end consumer spending annually.

7.6.2 Historical Disposable Personal Income (1999 to 2019)

From Federal Reserve Board data (refer to Appendix 7.5.2)

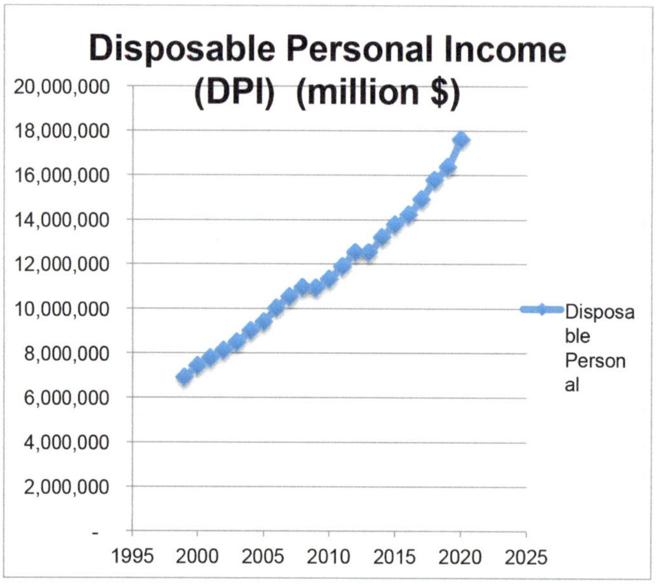

Figure 7.6.1.1 Plot of Historical Disposable Income

7.6.3 DPI Per Earner

Using the total number of the workforce in subsection and the DOI of 2019, we calculate an approximate value of $ 114,200 per year.

An approximate value per capita is $ 55,800 per year.

For a working person , its ability to make purchases as a consumer is obviously a function of his / her income. That is why we are concerned with the DPI.

7.7 Price Indices

7.7.1 CPI – Consumer Price Index

The **Consumer Price Index** (**CPI**) is a measure that examines the weighted average of prices of a basket of consumer goods and services, such as transportation, food, and medical care.

It is calculated factoring price changes in each item of the pre-determined basket.

Bureau of Labor Statistics, U.S. Department of Labor, *The Economics Daily*, Consumer Price Index increased 1.7 percent for year ending September 2019 on the Internet at https://www.bls.gov/opub/ted/2019/cpi-increased-1-point-7-percent-for-year-ending-september-2019.htm (visited *January 04, 2021*).

Figure 7.7.1 shows the Twelve-month percent change in the CPI (2009 -2019) BEA

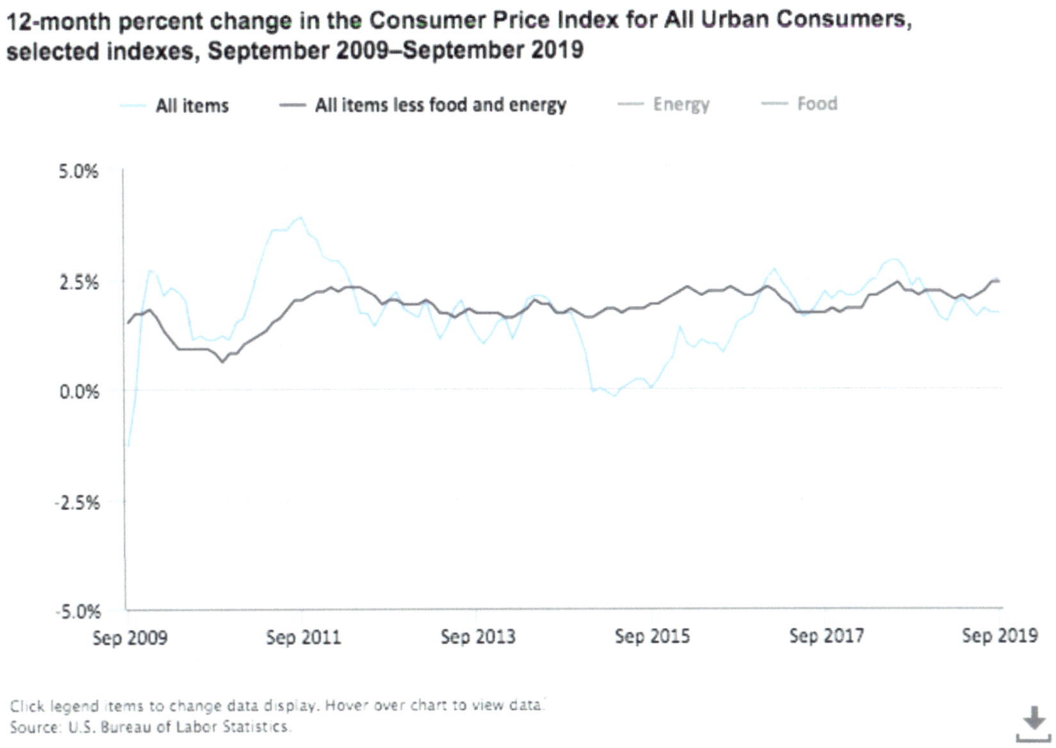

Figure 7.7.1 Twelve-month percent change in the CPI (2009 -2019) BEA

7.7.2 Producer Price Index

The producer price index (PPI), published by the Bureau of Labor Statistics (BLS), is a group of indices that calculates and represents the average movement in selling prices from domestic production over time.

- § The PPI is different from the CPI in that it measures costs from the viewpoint of industries that make the products whereas the CPI measure prices from the perspective of consumers.
- § The BLS separates PPI data into three main areas of classification namely industry, commodity, and commodity-based final and intermediate demand (FD-ID).
- § The PPI is considered an objective tool for adjusting prices in long-term purchasing agreements.

Please refer to subsection 7.4.10 containing representative PPI values.

7.8 Competition and Monopoly

For any country, the Government needs to allow enterprises to compete in a fairway in order to assure the consumer a competitive price. This is not an easy matter, Governments should avoid collusion of companies to engage in illegal behaviors in tenders whether public or even in tenders announced by private enterprises. This is a must for healthy economy that functions in accordance with the laws that are derived from the constitution that aims for the wellbeing of the citizens of the country.

It is not in the interest of the country to give a single company monopoly of a particular product or service. This is becoming more complex in latest years due to trends in consolidations/acquisitions of companies as obviously the large scale of operation gives economy of scale and minimizes duplications of functions (particularly indirect / corporate ones) leading in principle to lower costs for the production.

For enterprises that engage in production outside its home country there are the issues of the host country requirements. In some cases, the issues are regional (e.g. Euro. Zone).

In some instances, the protection of sensitive technology gets in the way of acceptance of a merger or an acquisition.

7.9 Government and the Economy

7.9.1 Government Debt

Refer to subsection 7.5.2

7.9.2 Yield Curve

Please see an example in Table 7.9.2.1

Table 7.9.2.1 Example of Daily Yield Curve CMT Rates

Daily Treasury Yield Curve CMT Rates	
	12/28/20
1 Month	0.09
2 Month	0.10
3 Month	0.11
6 Month	0.11
1 Year	0.11
2 Year	0.13
3 Year	0.17
5 Year	0.38
7 Year	0.65
10 Year	0.94
20 Year	1.46
30 Year	1.67

Of course, Government has a major influence on the economy – and making sure it moves in the optimum direction amid of the complex continuously varying conditions.

7.10 The Future; "Effect on the Labor Force"

In many instances, one sees concerns expressed that many persons in the work force in a country are threatened in their wellbeing due to advances in technology. They see the ever-increasing automation and in some cases items such as robots replacing humans on the job floor. In some cases. they could be talking with emotions because "their livelihood that is at stake". They may not understand / or appreciate the economics of handling some tasks with high precision or quick turnaround. On the other-hand, the fast pace of changes in the various industries require all employees to look out for themselves and acquire needed skills for tomorrows work place. They should look ahead. The employer has also an important role "to induce the work force to be ready for immediate and long-term futures". To follow the developments in the pertinent knowledge for jobs is a must for all employees and employers.

7.11 International Trade

An example of the BEA issued reports for the USA Trade is reproduced here to show the items included in the report.

The report is issued on June 4, 2020 and it appears to include some effects of the COVID 19 influence. The Figure 7.11.1 below shows graphically the trade deficit for the goods and services.

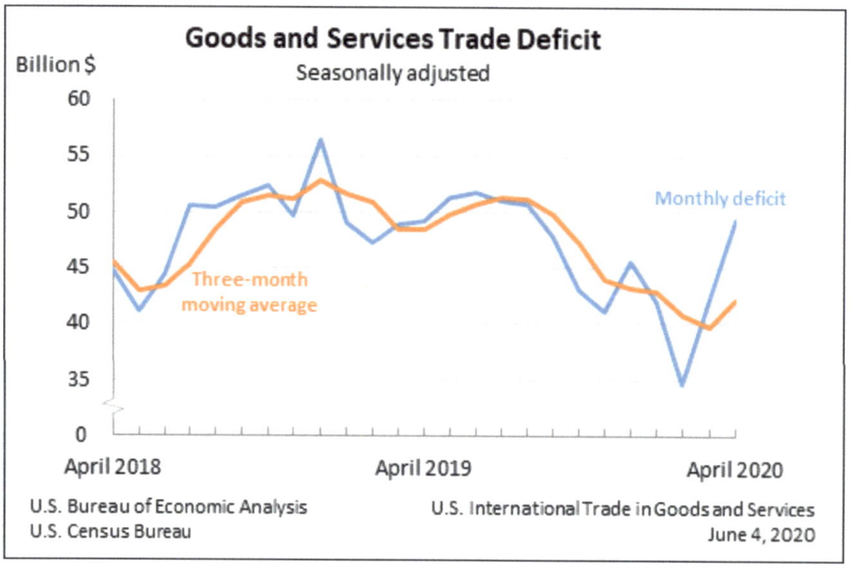

U.S. International Trade in Goods and Services, April 2020

Figure 7.11.1 Goods & Services Trade Deficit two year period April 2018- April 2020

Exhibits 19 and 19b in the "U.S. International Trade in Goods and Services, Annual Revision" report and exhibits 20 and 20b in the "U.S. International Trade in Goods and Services, April 2020" report were updated on June 8, 2020, to incorporate seasonally adjusted services data by geography that were not available with the initial publication. The release text and relevant tables in "U.S. Trade in Goods and Services by Selected Countries and Areas, 1999-present" were also updated.

The U.S. Census Bureau and the U.S. Bureau of Economic Analysis announced today that the **goods and services deficit** was $49.4 billion in April, up $7.1 billion from $42.3 billion in March, revised.

U.S. International Trade in Goods and Services Deficit

Deficit:	$49.4 Billion	+16.7%°
Exports:	$151.3 Billion	-20.5%°
Imports:	$200.7 Billion	-13.7%°

Coronavirus (COVID-19) Impact on April 2020 International Trade in Goods and Services

The declines in exports and imports that continued in April were, in part, due to the impact of COVID-19, as many businesses were operating at limited capacity or ceased operations completely, and the movement of travelers across borders was restricted. The full economic effects of the COVID-19 pandemic cannot be quantified in the trade statistics for April because the impacts are generally embedded in source data and cannot be separately identified. The Census Bureau and the Bureau of Economic Analysis have monitored data quality and determined estimates in this release meet publication standards. For more information on the impact of COVID-19 on the statistics, see the frequently asked questions on goods from the Census Bureau and on services from BEA.

Exports, Imports, and Balance (exhibit 1)[1]

April exports were $151.3 billion, $38.9 billion less than March exports. April imports were $200.7 billion, $31.8 billion less than March imports.

The April increase in the goods and services deficit reflected an increase in the goods deficit of $5.8 billion to $71.8 billion and a decrease in the services surplus of $1.3 billion to $22.4 billion.

Year-to-date, the goods and services deficit decreased $26.0 billion, or 13.4 percent, from the same period in 2019. Exports decreased $79.8 billion or 9.5 percent. Imports decreased $105.8 billion or 10.2 percent.

Three-Month Moving Averages (exhibit 2)

The average goods and services deficit increased $2.5 billion to $42.1 billion for the three months ending in April.

- Average exports decreased $19.7 billion to $184.4 billion in April.
- Average imports decreased $17.3 billion to $226.5 billion in April.

Year-over-year, the average goods and services deficit decreased $6.3 billion from the three months ending in April 2019.

- Average exports decreased $26.7 billion from April 2019.
- Average imports decreased $33.0 billion from April 2019.

Exports (exhibits 3, 6, and 7)

Exports of goods decreased $32.2 billion to $95.5 billion in April.

Exports of goods on a Census basis decreased $32.0 billion.

- Capital goods

  $10.1 billion.
  - Civilian aircraft engines decreased $2.2 billion.
  - Civilian aircraft decreased $2.1 billion.
  - Other industrial machinery decreased $1.0 billion.
- Industrial supplies and materials decreased $9.1 billion.
  - Crude oil decreased $2.0 billion.
  - Fuel oil decreased $1.7 billion.
  - Other petroleum products decreased $1.1 billion.
- Automotive vehicles, parts, and engines decreased $7.4 billion.
  - Automotive parts and accessories decreased $2.6 billion.
  - Passenger cars decreased $2.4 billion.
  - Trucks, buses, and special purpose vehicles decreased $1.2 billion.

Net balance of payments adjustments decreased $0.2 billion.

Exports of services decreased $6.7 billion to $55.8 billion in April.

- Travel decreased $3.4 billion.
- Transport decreased $2.3 billion.

Imports (exhibits 4, 6, and 8)

Imports of goods decreased $26.4 billion to $167.4 billion in April.

Imports of goods on a Census basis decreased $26.2 billion.

- Automotive vehicles, parts, and engines decreased $14.6 billion.
    - Passenger cars decreased $6.9 billion.
    - Automotive parts and accessories decreased $4.1 billion.
    - Trucks, buses, and special purpose vehicles decreased $2.3 billion.
- Capital goods decreased $5.8 billion.
    - Semiconductors decreased $1.4 billion.
    - Electric apparatus decreased $0.8 billion.
    - Civilian aircraft decreased $0.7 billion.
    - Civilian aircraft engines decreased $0.7 billion.
- Consumer goods decreased $3.1 billion.
    - Pharmaceutical preparations decreased $1.2 billion.
    - Cotton apparel and household goods decreased $1.1 billion.

Net balance of payments adjustments decreased $0.2 billion.

Imports of services decreased $5.4 billion to $33.3 billion in April.

- Travel decreased $2.9 billion.
- Transport decreased $2.0 billion.

Real Goods in 2012 Dollars – Census Basis (exhibit 11)

The real goods deficit increased $4.1 billion to $80.0 billion in April.

- Real exports of goods decreased $29.2 billion to $113.8 billion.
- Real imports of goods decreased $25.1 billion to $193.8 billion.

Revisions

Exports and imports of goods and services for all months through March 2020 shown in this release reflect the incorporation of annual revisions to the goods and services series. See the "Notice" in this release for a description of the revisions.

Revisions to March exports

- Exports of goods were revised down $0.4 billion.
- Exports of services were revised up $2.8 billion.

Revisions to March imports

- Imports of goods were revised up less than $0.1 billion.
- Imports of services were revised up $0.3 billion.

Goods by Selected Countries and Areas: Monthly – Census Basis (exhibit 19)

The April figures show surpluses, in billions of dollars, with South and Central America ($2.9), OPEC ($1.4), Brazil ($0.8), Saudi Arabia ($0.3), and Hong Kong (less than $0.1). Deficits were recorded, in billions of dollars, with China ($26.0), European Union ($14.3), Germany ($4.0), Japan ($3.6), Mexico ($3.3), South Korea ($2.3), Taiwan ($2.2), Italy ($2.0), India ($1.9), France ($1.4), Singapore ($1.1), Canada ($0.4), and United Kingdom ($0.4).

- The deficit with China increased $9.0 billion to $26.0 billion in April. Exports increased $2.1 billion to $9.3 billion, and imports increased $11.0 billion to $35.2 billion.
- The surplus with South and Central America decreased $2.1 billion to $2.9 billion in April. Exports decreased $4.3 billion to $8.7 billion, and imports decreased $2.2 billion to $5.8 billion.
- The deficit with Mexico decreased $5.6 billion to $3.3 billion in April. Exports decreased $7.0 billion to $12.4 billion, and imports decreased $12.6 billion to $15.8 billion.

Goods and Services by Selected Countries and Areas: Quarterly – Balance of Payments Basis (exhibit 20)

Statistics on trade in goods and services by country and area are only available quarterly, with one-month lag. With this release, first-quarter figures are now available.

The first-quarter figures show surpluses, in billions of dollars, with South and Central America ($24.0), OPEC ($9.4), Brazil ($9.0), Hong Kong ($6.0), United Kingdom ($5.4), Singapore ($4.1), and Saudi Arabia ($3.1). Deficits were recorded, in billions of dollars, with China ($53.7), European Union ($29.8), Mexico ($28.1), Germany ($15.9), Japan ($12.2), Italy ($8.0), India ($6.4), Taiwan ($4.1), France ($1.1), South Korea ($0.9), and Canada ($0.3).

- The deficit with China decreased $14.3 billion to $53.7 billion in the first quarter. Exports decreased $4.4 billion to $35.0 billion and imports decreased $18.7 billion to $88.7 billion.
- The deficit with Canada decreased $3.0 billion to $0.3 billion in the first quarter. Exports decreased $2.4 billion to $86.7 billion and imports decreased $5.5 billion to $87.0 billion.
- The deficit with Mexico increased $1.4 billion to $28.1 billion in the first quarter. Exports decreased $0.8 billion to $69.2 billion and imports increased $0.6 billion to $97.3 billion.

* * *

All statistics referenced are seasonally adjusted; statistics are on a balance of payments basis unless otherwise specified. Additional statistics, including not seasonally adjusted statistics and details for goods on a Census basis, are available in exhibits 1-20b of this release. For information on data sources, definitions, and revision procedures, see the explanatory notes in this release. The full release can be found at www.census.gov/foreign-trade/Press-Release/current_press_release/index.html or www.bea.gov/data/intl-trade-investment/international-trade-goods-and-services. The full schedule is available in the Census Bureau's Economic Briefing Room at www.census.gov/economic-indicators/ or on BEA's website at www.bea.gov/news/schedule.

7.12 Currency & Exchange Rate

For any country, the balance of its trade has very significant effect on its currency. I) A country that has goods and services other countries are interested in buying will command a high value for its currency (demand driven – per subsection 7.1). II) Contrarily, a country that has fewer products and services that other countries are not eager to buy, will suffer. The equation of the export posture when the export of the country exceeds the imports will lead to I). In the other side when the exports are less than the imports, we shall see II).

We see countries that have high technological products and services have strong currency. Also, on the energy side we see many Middle Eastern countries that export energy (fuels – be LNG, natural gas or oil) also have good possibility to peg their currency to the USD. The exchange rate is closely related to the combination of the knowledge and energy endowments of the country.

In the last years, countries enter in agreements either bilateral or multilateral/regional agreements with other countries affect imports and exports.

Political events at any time have effects on trade. Major trade transactions could have impact on the exchange rate of the currency of a country.

An example of the historical variation of the USD and the Euro is included as an example hereafter.

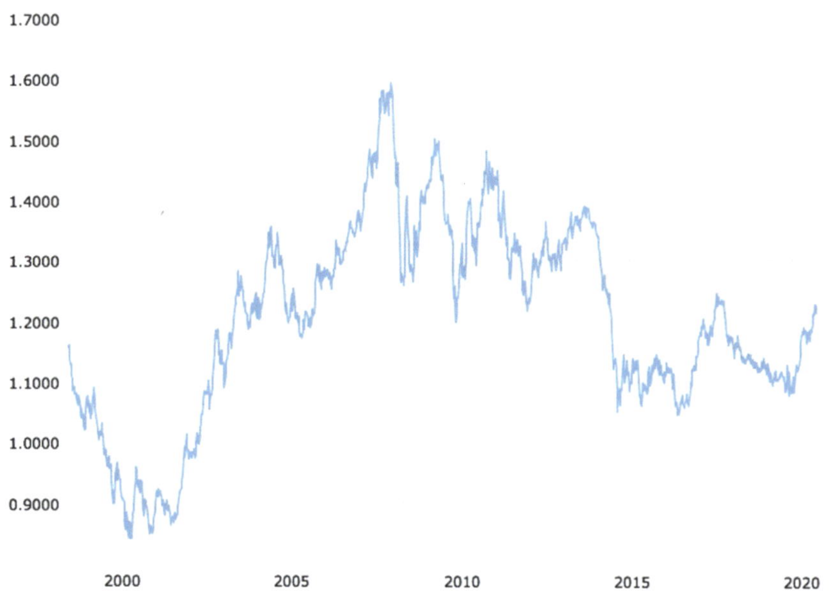

Figure 7.12.1 Euro – USD exchange rate (2000 -2020)

Source: https://www.macrotrends.net/2548/euro-dollar-exchange-rate-historical-chart

7.13 Conflicts, Sanctions, Embargos, Wars

Since the end of the second-world war the world did not see a massive war but there were plenty of conflicts that took place across four continents (Asia, Africa, Europe and South America). Many regions saw long wars that lasted decades; in some instances, the short wars did not settle the conflicts. The cease-fire or the impasse ended up in long-protracted suffering to locals, resulting from imposition of sanctions and embargos. Many of the countries affected by conflicts could not pursue development plans and their economy was negatively impacted,; in some cases, catastrophic deterioration in living conditions was seen.

With respect to the theme of this book, in the unsettling situations that erupt from conflicts, both knowledge and energy are negatively impacted and the result of low growth if not negative growth is encountered.

7.14 Lessons Learned from the COVID 19 Epidemic of 2020

Since early 2020 the COVID 19 epidemic has had major effects on life. To stop / minimize the spread of the virus governments instituted measures based on the recommendations of the WHO (World Health Organizations). Because the virus was new, it took many countries time to learn and try different ways to avoid the transmittal of the virus. In the early months of 2020, the spread had high rates, hospitals and health care providers were overburdened and the fatalities from the virus were high. Efforts to cope with the COVID 19 cases in hospitals and health institutions were handicapped with many limitations on available medical protective gear for heath personnel. Dealing with a new virus with many unknown features proved challenging to both the professionals / experts and to politicians / government officials.

Securing the necessary medical and protective in a short time amid high demand led to very high prices. Companies that were never engaged in manufacturing the needed protective items and medical equipment had to step in and contributed to solving the shortages.

In the meantime, unfortunately many companies engaged in unprofessional behavior asking for exorbitantly high prices. Some countries had to import such items and pay very high bills.

To curb the high rates of infections measures of social distancing, mask wearing, plus hand cleaning were required in some cases enforced. Some countries were under local governments oversight.

Amid the fast changes that were encountered day in day out, the enterprises, the public and governments were learning. Sharing of information / equipment and health professionals across countries and from one locality to another proved extremely useful.

It was necessary to minimize the distance between personnel in the workplace and schools to curtail the transmittal between peers coexisting indoors in a confined space. Scheduling 50% of the work force or more to work from home became commonplace.

Working from home was forced on many of us, but thanks to technologies that were already in place with some refinements that were introduced by many companies offering teleconferencing services, many business meetings were conducted and a good level of communications was reached.

In some instances when the situation in a locality deteriorated to a high degree some establishments were closed. Many small and medium size enterprises had to shut down, and many employees were laid off. Unemployment skyrocketed and reached unprecedented levels. This led governments with the help of legislative bodies to introduce stimulus packages to help the citizens that were affected.

Some sectors in the economy were hurt very badly e.g., travel and leisure because of the restrictions and fear of the public on being confined in space.

The stimulus packages were essential in propping the economies in many countries, but the long-term effect remains to be seen.

The massive efforts to develop vaccines were quite impressive and at the time of finalizing this book (Nov. 2021), three vaccines have gained certifications in the USA and Europe and large percentages of the populations have been vaccinated.. In parallel efforts continued in many countries to develop anti COVID 19 vaccines.

The difficult days that people suffered over 2020 will have profound social effects on parents and children. The treated COVID 19 patients still could see unknown effects on their long-term health.

Knowledge was a saver in coping with COVID 19. Further, countries that had in place adequate preparedness fared better than others. The shutting down of the economies or at least some areas caused drastic negative impacts. It is not known how and when economies will recover.

7.15 Analysis of Industrial Enterprises

It is vital that before introducing the new view of the knowledge & energy in subsection 9.1 (building on subsection 5.1) to review what the current practice is.

As noted in subsection 6.2 historical data should be the starting point for the development of the K&E model. Without getting data from a particular enterprise, the author decided to gather information from various US Federal Government agencies. The author wishes to acknowledge the excellent data resources of the entities named below:

Referring to sections 5 and 6, the K&E model was introduced and applied to a number of sample manufacturing and services enterprises. In this subsection available data for some industries are reviewed to complement the contents of sections 5 and 6. Further, the contents in subsections 7.3, "Energy", 7.4 "Labor", 7.7.2 "PPI", supplement the following information. Again, underlining the intention is to substantiate the roles of knowledge and energy in the economy of the enterprises.

- US Department of Commerce
  - Bureau of Economic Analysis "BEA"
- The US Department of the Treasury

- Department of Energy
  - International Energy Agency "IEA"
- Department of Labor
  - The US Bureau of Labor Statistics
- Federal Reserve Board
- Central Intelligence Agency "CIA"

7.15.1 Industry Classification & Groups

The information in this section follows the tables of industries that BEA has adopted.

The author filtered groupings of the industries manufacturing and services. The detailed grouping had 1064 rows. The "groupings with four digits" had 156 rows. The "groupings with three digits" had 53 rows. The "additional" has 13 rows. These do not have digital numbers but have characters designations for the industry items.

7.15.2 Industry Input & Output

A basic understanding of an industry's current economic activity and how it relates to the total U.S. economy can be obtained from the two main tables in the I-O accounts—the standard makes use tables. These two tables provide consistent statistics on an industry's production and its relationship with other industries for a given year. They offer a wealth of information about the size of the U.S economy, the relative size of specific industries, what and how much is produced by specific industries, the technology used by specific industries, the incomes generated by production, and the size and scope of an industry's market.

The relevance of the information / data was clearly made in the applications of the K&E model in sections 5 and 6.

The origin of Table 7.15.2 is the "Industry-by-Industry Total Requirements. After Redefinitions" – 1997- 2019 the requirements for each year 1997 to 2019. Each year has a matrix of 71x71 columns x rows containing cross-reference of the industries.

For the demonstration of some points, five industries were selected, whose rows are marked in colors.

i) 221 Utilities

ii) 23 Construction
iii) 335 Electrical Equipment, Appliances and components
iv) 523 Securities, Commodities Contracts, and investments
v) 541 Computer systems and related services

Table 7.15.2.1 shows the values corresponding to the above five groups for the year 2019 as extracted

Table 7.15.2.1 Values corresponding to the selected five industry groups for **year 2019**

2019

	IOCode	22	23	335	523	5415
Industries/ Industries	Name	Utilities	Construction	Electrical equipment, appliances, and components	Securities, commodity contracts, and investments	Computer systems design and related services
111CA	Farms	0.0232613	0.0156177	0.0109226	0.0063855	0.0073299
113FF	Forestry, fishing, and related activities	0.0042866	0.0025877	0.0034905	0.0025242	0.0028231
211	Oil and gas extraction	0.0232752	0.0088835	0.0051224	0.0096368	0.0390969
212	Mining, except oil and gas	0.0312946	0.0240750	0.0063587	0.0271659	0.0116478
213	Support activities for mining	0.0044258	0.0087391	0.0063032	0.0628590	0.0052998
22	Utilities	1.0406396	0.0206265	0.0020894	0.0056157	0.0083866
23	Construction	0.0107417	1.0053979	0.0279653	0.0058646	0.0083836
321	Wood products	0.0187632	0.0072646	0.0226181	0.0051617	0.0083665
327	Nonmetallic mineral products	0.0256832	0.0096835	0.0046251	0.0079321	0.0085428
331	Primary metals	0.0307728	0.0110829	0.0164955	0.0142537	0.0102555
332	Fabricated metal products	0.0198377	0.0079190	0.0149203	0.0118467	0.0123926
333	Machinery	0.0145815	0.0062407	0.0546868	0.0111912	0.0161351
334	Computer and electronic products	0.0045434	0.0016999	0.0068966	0.0024126	0.0045176
335	Electrical equipment, appliances, and components	0.0134426	0.0054428	1.0867194	0.0076373	0.0069000
3361MV	Motor vehicles, bodies and trailers, and parts	0.0169912	0.0068740	0.0264015	0.0094523	0.0102202
3364OT	Other transportation equipment	0.0087228	0.0037345	0.0154710	0.0071988	0.0104436

337	Furniture and related products	0.0175866	0.0076896	0.0083032	0.0124433	0.0099234
339	Miscellaneous manufacturing	0.0120601	0.0051538	0.0124473	0.0096285	0.0087852
311FT	Food and beverage and tobacco products	0.0208387	0.0101054	0.0070835	0.0067345	0.0092386
313TT	Textile mills and textile product mills	0.0273973	0.0086680	0.0055420	0.0055591	0.0092066
315AL	Apparel and leather and allied products	0.0136615	0.0056653	0.0033294	0.0056555	0.0092257
322	Paper products	0.0309402	0.0111405	0.0085622	0.0057684	0.0089938
323	Printing and related support activities	0.0226253	0.0094379	0.0062515	0.0054548	0.0127320
324	Petroleum and coal products	0.0242480	0.0207882	0.0041683	0.0073338	0.0251506
325	Chemical products	0.0242495	0.0079744	0.0052611	0.0048309	0.0085416
326	Plastics and rubber products	0.0267461	0.0089822	0.0108564	0.0059333	0.0094184
42	Wholesale trade	0.0134740	0.0078901	0.0021097	0.0068216	0.0136289
441	Motor vehicle and parts dealers	0.0236108	0.0119925	0.0036553	0.0057906	0.0194471
445	Food and beverage stores	0.0388018	0.0138297	0.0024421	0.0061720	0.0081145
452	General merchandise stores	0.0271555	0.0110679	0.0029500	0.0067897	0.0115987
4A0	Other retail	0.0244734	0.0145250	0.0028759	0.0070021	0.0112741
481	Air transportation	0.0057765	0.0062192	0.0016075	0.0293939	0.0075049
482	Rail transportation	0.0069650	0.0294410	0.0053410	0.0198548	0.0223728
483	Water transportation	0.0105291	0.0081438	0.0040814	0.1840007	0.0133017
484	Truck transportation	0.0119367	0.0090692	0.0028116	0.0118062	0.0111022
485	Transit and ground passenger transportation	0.0081806	0.0128904	0.0077565	0.0384647	0.0096786
486	Pipeline transportation	0.0056537	0.0074863	0.0009935	0.0027695	0.0018767
487OS	Other transportation and support activities	0.0092264	0.0130085	0.0035837	0.0135503	0.0082102
493	Warehousing and storage	0.0663876	0.0207614	0.0074957	0.0058551	0.0075861
511	Publishing industries, except internet (includes software)	0.0039877	0.0023100	0.0010876	0.0034824	0.0215342
512	Motion picture and sound recording industries	0.0081367	0.0057728	0.0015825	0.0082267	0.0105551

513	Broadcasting and telecommunications	0.0086805	0.0073637	0.0110328	0.0049325	0.0158850
514	Data processing, internet publishing, and other information services	0.0087594	0.0054553	0.0036078	0.0046119	0.0329936
521CI	Federal Reserve banks, credit intermediation, and related activities	0.0073634	0.0076485	0.0014797	0.0474731	0.0145168
523	Securities, commodity contracts, and investments	0.0085868	0.0090606	0.0020004	1.1116783	0.0235982
524	Insurance carriers and related activities	0.0077001	0.0076858	0.0012374	0.0250237	0.0087543
525	Funds, trusts, and other financial vehicles	0.0129082	0.0088692	0.0018011	0.7104165	0.0200472
HS	Housing	0.0012924	0.0333438	0.0011038	0.0046614	0.0014423
ORE	Other real estate	0.0678283	0.0714424	0.0056767	0.0150000	0.0098745
532RL	Rental and leasing services and lessors of intangible assets	0.0079438	0.0052955	0.0021330	0.0102456	0.0065355
5411	Legal services	0.0072358	0.0066893	0.0012079	0.0066668	0.0160326
5415	Computer systems design and related services	0.0026707	0.0019245	0.0027798	0.0031142	1.0140689
5412OP	Miscellaneous professional, scientific, and technical services	0.0081035	0.0064634	0.0046706	0.0068237	0.0148892
55	Management of companies and enterprises	0.0177140	0.0080886	0.0020519	0.0231894	0.0400690
561	Administrative and support services	0.0087047	0.0056822	0.0049889	0.0082955	0.0208754
562	Waste management and remediation services	0.0094270	0.0053025	0.0125899	0.0058512	0.0091441
61	Educational services	0.0207351	0.0095265	0.0039798	0.0045811	0.0094090
621	Ambulatory health care services	0.0082349	0.0056416	0.0016438	0.0057876	0.0098646
622	Hospitals	0.0142924	0.0079737	0.0028943	0.0197737	0.0185418
623	Nursing and residential care facilities	0.0171926	0.0116657	0.0058981	0.0076441	0.0106588
624	Social assistance	0.0131141	0.0113682	0.0043192	0.0155479	0.0104405
711AS	Performing arts, spectator sports, museums, and related activities	0.0091965	0.0071384	0.0017313	0.0060504	0.0053680

713	Amusements, gambling, and recreation industries	0.0197588	0.0117592	0.0041910	0.0060187	0.0126302
721	Accommodation	0.0263492	0.0092978	0.0036225	0.0045850	0.0078341
722	Food services and drinking places	0.0281722	0.0127322	0.0030491	0.0148955	0.0099944
81	Other services, except government	0.0106223	0.0103219	0.0065061	0.0252271	0.0090010
GFGD	Federal general government (defense)	0.0057854	0.0149769	0.0047892	0.0031362	0.0393249
GFGN	Federal general government (nondefense)	0.0042675	0.0065952	0.0041520	0.0062590	0.0285986
GFE	Federal government enterprises	0.0123116	0.0134226	0.0028132	0.0074684	0.0089899
GSLG	State and local general government	0.0125702	0.0177368	0.0030943	0.0035693	0.0129504
GSLE	State and local government enterprises	0.0121866	0.1556014	0.0142209	0.0147597	0.0151766

Table 7.15.2.2 Values corresponding to the selected five groups for **year 2010**

2010

IOCode	22	23	335	523	5415
Name	Utilities	Construction	Electrical equipment, appliances, and components	Securities, commodity contracts, and investments	Computer systems design and related services
Farms	0.0343114	0.0173515	0.0055927	0.0078793	0.0054293
Forestry, fishing, and related activities	0.0110122	0.0044057	0.0077860	0.0043995	0.0034868
Oil and gas extraction	0.0217103	0.0123275	0.0026951	0.0076637	0.0108366
Mining, except oil and gas	0.0416715	0.0248424	0.0039926	0.0184890	0.0055825
Support activities for mining	0.0074796	0.0180160	0.0044316	0.0449708	0.0046316
Utilities	1.0647900	0.0224005	0.0019041	0.0071090	0.0074357
Construction	0.0166403	1.0058500	0.0255288	0.0077485	0.0066429
Wood products	0.0387359	0.0102774	0.0127131	0.0077109	0.0071753
Nonmetallic mineral products	0.0554139	0.0140608	0.0059061	0.0099659	0.0074989
Primary metals	0.0558410	0.0153897	0.0187260	0.0315346	0.0082283
Fabricated metal products	0.0355306	0.0104550	0.0137958	0.0294354	0.0092375

Machinery	0.0252297	0.0081864	0.0444288	0.0253953	0.0105116
Computer and electronic products	0.0116653	0.0036251	0.0145802	0.0065765	0.0092293
Electrical equipment, appliances, and components	0.0261215	0.0077245	**1.0763400**	0.0187814	0.0058673
Motor vehicles, bodies and trailers, and parts	0.0324767	0.0093392	0.0238866	0.0211591	0.0086868
Other transportation equipment	0.0185852	0.0058587	0.0172617	0.0134382	0.0123016
Furniture and related products	0.0302265	0.0097354	0.0064503	0.0274383	0.0082442
Miscellaneous manufacturing	0.0217301	0.0070988	0.0105560	0.0265547	0.0080495
Food and beverage and tobacco products	0.0346155	0.0119741	0.0049420	0.0081441	0.0068502
Textile mills and textile product mills	0.0433559	0.0103274	0.0046302	0.0070536	0.0072581
Apparel and leather and allied products	0.0238112	0.0078523	0.0027851	0.0067448	0.0078235
Paper products	0.0541730	0.0129140	0.0064917	0.0070243	0.0065804
Printing and related support activities	0.0303972	0.0102744	0.0045738	0.0061191	0.0091221
Petroleum and coal products	0.0217391	0.0196233	0.0026316	0.0063492	0.0088472
Chemical products	0.0376281	0.0100058	0.0043594	0.0059595	0.0065833
Plastics and rubber products	0.0418960	0.0109301	0.0090121	0.0075980	0.0074319
Wholesale trade	0.0170175	0.0064952	0.0016982	0.0064652	0.0095309
Motor vehicle and parts dealers	0.0282305	0.0098744	0.0025323	0.0051376	0.0113729
Food and beverage stores	0.0546552	0.0121522	0.0018751	0.0060186	0.0059508
General merchandise stores	0.0420503	0.0108949	0.0030974	0.0071414	0.0093859
Other retail	0.0302813	0.0115695	0.0019556	0.0064747	0.0075418
Air transportation	0.0113204	0.0073956	0.0019805	0.0171309	0.0086550
Rail transportation	0.0107700	0.0299218	0.0049165	0.0239176	0.0182974
Water transportation	0.0154926	0.0093971	0.0040475	0.1902860	0.0117894
Truck transportation	0.0194154	0.0104312	0.0022964	0.0092834	0.0062154
Transit and ground passenger transportation	0.0094373	0.0071167	0.0019994	0.0210776	0.0037421

Pipeline transportation	0.0547261	0.0400216	0.0027394	0.0076312	0.0036844
Other transportation and support activities	0.0321997	0.0117123	0.0021470	0.0145468	0.0041648
Warehousing and storage	0.1103760	0.0234748	0.0066294	0.0061458	0.0060417
Publishing industries, except internet (includes software)	0.0090011	0.0041288	0.0016089	0.0071016	0.0191660
Motion picture and sound recording industries	0.0066256	0.0035840	0.0008207	0.0050313	0.0041720
Broadcasting and telecommunications	0.0107636	0.0079756	0.0083622	0.0047671	0.0105670
Data processing, internet publishing, and other information services	0.0128975	0.0050453	0.0038782	0.0045688	0.0281022
Federal Reserve banks, credit intermediation, and related activities	0.0108504	0.0107063	0.0015743	0.0574801	0.0169452
Securities, commodity contracts, and investments	0.0115497	0.0096111	0.0020625	**1.1157800**	0.0259422
Insurance carriers and related activities	0.0046322	0.0042989	0.0005806	0.0163157	0.0038693
Funds, trusts, and other financial vehicles	0.0216001	0.0111496	0.0019682	0.7088500	0.0223048
Housing	0.0014916	0.0376368	0.0010511	0.0049266	0.0012748
Other real estate	0.0793588	0.0722783	0.0038777	0.0131052	0.0073144
Rental and leasing services and lessors of intangible assets	0.0151638	0.0067223	0.0019660	0.0141696	0.0071612
Legal services	0.0090928	0.0064674	0.0011516	0.0089854	0.0168767
Computer systems design and related services	0.0065253	0.0035593	0.0040461	0.0053362	**1.0181800**
Miscellaneous professional, scientific, and technical services	0.0110913	0.0069854	0.0042255	0.0067925	0.0118973
Management of companies and enterprises	0.0246603	0.0073893	0.0020444	0.0285185	0.0306427
Administrative and support services	0.0101395	0.0050950	0.0043063	0.0081049	0.0145254

Waste management and remediation services	0.0135227	0.0057661	0.0090135	0.0065158	0.0063753
Educational services	0.0247589	0.0095683	0.0038199	0.0048294	0.0077742
Ambulatory health care services	0.0132999	0.0072614	0.0015641	0.0066614	0.0078633
Hospitals	0.0197180	0.0077592	0.0024636	0.0138441	0.0119907
Nursing and residential care facilities	0.0244575	0.0130328	0.0037739	0.0079764	0.0083031
Social assistance	0.0232735	0.0152181	0.0042056	0.0161401	0.0093576
Performing arts, spectator sports, museums, and related activities	0.0153117	0.0087336	0.0014235	0.0091813	0.0047679
Amusements, gambling, and recreation industries	0.0302828	0.0130848	0.0026385	0.0056798	0.0077621
Accommodation	0.0578957	0.0160951	0.0032277	0.0064108	0.0076182
Food services and drinking places	0.0437438	0.0125996	0.0027571	0.0147707	0.0079721
Other services, except government	0.0174851	0.0144810	0.0055627	0.0218534	0.0065552
Federal general government (defense)	0.0107710	0.0128734	0.0046378	0.0042482	0.0460428
Federal general government (nondefense)	0.0100423	0.0085677	0.0092437	0.0040876	0.0579588
Federal government enterprises	0.0203700	0.0184167	0.0025528	0.0115730	0.0097352
State and local general government	0.0170137	0.0175899	0.0029474	0.0036749	0.0089599
State and local government enterprises	0.0192718	0.1501260	0.0130093	0.0149320	0.0112111

In the above two tables, one can see the value shown in bold red is close to 1 signifying the main activities for the group; the other industries give the contributions when the value is with 0.00 or .000 meaning negligible contribution. The essence of the contributions is the indirect contribution to the group. This is in line with the schematics shown in sections 6.1.

When one compares the values of 2019 to those of 2010, one can recognize the trend in outsourcing or the opposite.

7.15.3 The Use of Commodities by Industries (1997 – 2019)

Complementing the information in subsection 7.15.2 we include the data in 7.15.3 giving the matrix of 25 items (16 industries and 9 supplementary categories), the data are in $ for years 1997 to 2019 (each in a separate tab).

Table 7.15.3 contains extracts for three industry groups: utilities, construction and finance, insurance, real-estate rental and leasing, and information. One can see the contributions of the other industry groups to the specific whose own value is shown in ball red. The columns T005 is for the totals (of the industry groups).

Table 7.15.3 Commodities / Industries for 2019 (Values in million $ - 2019)

	IOCode	T005	22	23	FIRE	51
Commodities/ Industries	Name	Total Intermediate	Utilities	Construction	Finance, insurance, real estate, rental, and leasing	Information
11	Agriculture, forestry, fishing, and hunting	277,054	4,401	1,681	33,682	582
21	Mining	270,466	9,069	3,501	42,756	2,180
22	Utilities	167,962	22,602	6,840	13,517	2,809
23	Construction	797,302	4,753	217	56,687	8,584
31G	Manufacturing	3,913,382	55,663	14,197	108,329	22,650
42	Wholesale trade	829,392	15,113	1,983	189,839	25,313
44RT	Retail trade	792,247	33,156	4,465	234,543	29,575
48TW	Transportation and warehousing	656,413	12,237	5,923	138,470	13,858
51	Information	813,674	5,448	2,456	89,086	237,039
FIRE	Finance, insurance, real estate, rental, and leasing	2,751,366	108,576	160,480	1,420,134	73,613

PROF	**Professional and business services**	**1,566,085**	16,373	1,799	301,006	121,163
6	**Educational services, health care, and social assistance**	**1,143,502**	18,894	1,651	335,075	42,412
7	**Arts, entertainment, recreation, accommodation, and food services**	**661,377**	28,745	2,899	160,697	22,982
81	**Other services, except government**	**296,403**	3,963	3,498	86,867	18,944
G	**Government**	**1,447,135**	26,836	95,792	193,748	104,515
T001	**Total Intermediate**	**6,383,760**	**365,828**	**307,383**	**3,404,437**	**726,220**
F010	**Personal consumption expenditures**	**4,544,601**	288,088	...	3,627,344	602,362
F020	**Private fixed investment**	**3,702,122**	...	1,123,073	162,237	256,618
F030	**Change in private inventories**	**49,100**	...	...	...	1,079
F040	**Exports of goods and services**	**2,178,432**	4,375	101	235,679	84,436
F100	**Government consumption expenditures and gross investment**	**3,747,874**	...	356,640	...	13,421
T019	**Total use of products**	**0,605,889**	658,291	1,787,197	7,429,698	1,684,136

In Table 15.3.3 one can see in the row T001 "Total Intermediate" the values for the four selected industries. The last two are services.

Under T019, one can see the total use of products. F040 shows the export of goods and services. Row F100 shows the government consumption.

7.16 Economies of Developing Nations - Comments

In the previous subsections we used the data of US Economy being the leading economy of the world.

Of course, among the developed countries and in particular the OECD countries we can find robust economies as well. As noted earlier the idea behind sections 7 and 8, is to prepare for section 9 that takes the knowledge & energy model introduced in subsection 5.1 to higher level while relating to well-known concepts in economics and investments.

The author believes that the advances over the last four decades have led to a more well-off society today compared to decades ago. There could be variety of measures to this but the personal disposal income (discussed in subsection 7.6) could be used as a starting point. The distribution of the income, with less disparity among the workforce should be considered. Of course, the affordability of products and services is another key factor.

Let's turn now to the developing countries, many of which, unfortunately, are to attempting to develop their economies amid various political systems in these countries. The change in thinking of the developing countries that started in the early 1990's, that the poorer countries, if developed, would be better consumers with mutual-benefits through two-way trade rather than just getting raw materials from the poorer countries.

With increased trade with the developing countries came open channels to various aspects of life in the developed world. Over the last two decades many developing countries made progress – each according to its own circumstances. Many had to concentrate on developing infrastructure. Politics in some countries have hindered the growth rate. At any rate knowledge and energy are key to advancing the economies. Attention to education and training in all fields are keys to accelerating the advancements.

7.17 References & Bibliography

1) **Walters J. Wessels**, *Economics*, Fourth Edition, Business Review Books, Barron's, ISBN-13 978-0-7641-34197, ISBN-10 0-7641-3419-1
2) International Energy Agency https://www.iea.org
3) U.S. Bureau of Economic Analysis - https://www.bea.gov/data
4) U.S. Bureau of Labor Statistics -
5) https://www.bls.gov
6) U.S. Department of Treasury - https://home.treasury.gov

Section Eight: INVESTMENTS

The intention of this section is not to provide a complete description of investments – an important field with ample sources of information; the intent is to underline the interplay of knowledge and energy in how enterprises make investments.

In subsection 5.2, project phases were discussed. It was underlined that raising a significant amount of money necessary to fund the building of the facilities of the project is very important. At that stage the operations of the facilities normally are planned.

If external financing (getting loan(s)) from financial institutions is to be followed, then convincing an investor to share the equity and a financial institution to fund the debt is called for.

The goal is to establish how the assets required for the operational phase will be funded.

In this section we look at securing the financial support from the start of the development, through the construction phase, up to the commissioning and testing leading to the start of commercial operation.

During the operation phase – recovery of the investments in the fixed assets and profits are attained.

8.1 Financing -Terms

Investing is to allocate money in the expectation of some benefit in the future.

In finance, the benefit from an investment is called a return. The return may consist of a gain or a loss realized from the sale of an asset or an investment, unrealized capital appreciation (or depreciation), or investment income such as dividends, interest, rental income, or a combination of capital gain and income. The return may also include currency gains or losses due to changes in the foreign currency exchange rates.

Investors generally expect higher returns from riskier investments. When a low-risk investment is made, the return is also generally low.

It is noted that for financing the interest rate and the terms of financing are quite important. The project sponsor investigates what the interest rate is for the loan, whether variable or fixed, and the loan term (period for repayment). Please refer to subsections 7.5.3 "interest rate" and 7.5.4 "inflation rate". In the current era that is characterized by relatively low interest rate, the issue is not as critical as it was in the 1980s and 1990s. One should note the long-term interest rate is influenced by the perception in the market of the inflation over long horizon.

8.2 Financing Types

When an enterprise considers a project for setting up a new facility to produce a particular product or products, it generally would follow a phased approach to the one described in subsection 5.2. At the feasibility study stage, consideration is given on how project financing will be secured.

8.2.1 Self-financing

In this case, the enterprise has enough cash that it can finance the project phases until the facility is ready to start the production. The asset would have been completed from internal resources. This approach most probably is used for small projects.

8.2.2 Financing under Project Structure

Financing under project finance structure, generally, refers to securing the debt against long term agreement between the project owner and an off-taker. The principal, the enterprise decides it will finance all the equity of the project or it will solicit co-investors. Generally, the financing will be on an equity / debt (e/d) for the total capital of the project e.g., e/d can be 30/70.

Debt is obtained through loan from a financial institution, a bank or a group of banks for large projects. The financing institution uses ample criteria to be satisfied that the loan will be paid in full in accordance with the financing agreement. The process of arrangement of a loan requires experience within the enterprise, and if not enough internally, the enterprise gets support from an external financial advisor.

The financing package to be submitted to the financial institution must be substantiated and include many documents. A cornerstone is the proven experience of project participants starting from the sponsors (the equity holders) and their financial records. A newly formed company will face more scrutiny when applying to secure a loan for a project.

Please refer to the discussions found under subsection 8.4 “financial model”.

8.2.3 Financing under Merchant Structure

Referring to the example of project structure finance found under subsection 8.4 for technical & financial analysis a financial model is set up in a similar manner, however, different parameters are to be varied under different scenarios that assume different variants both on the expenses and the revenue. In many instances probabilistic techniques are applied. Also, the time horizon of the model prediction is only few years, or with more dependable results in the early years in comparison to the latter.

In case of Merchant structure, the debt/equity ratio is generally less than in the case of the financing of project finance (case in 8.2.2). The uncertainties inherent in merchant financing are obvious.

Management of enterprises engaged in merchant financing always revert to fundamental analysis results including historical data and market information.

8.3 Financing for Projects Abroad (Outside the Enterprise Home Country)

An enterprise engaged in an overseas investment must try to leverage its knowledge in expanding its operation beyond the home country. The knowledge of the conditions, laws, and political structure, and taxes in the host country is crucial to making an investment in the host country. The attitudes of customers in the host country must be studied and adapted to the products of local tastes. Any local content requirements must be factored. The opportunities for local financing should be assessed. Examination of any restrictions on repatriation of the brought in capital and the profits must be evaluated.

The enterprise must also investigate the time horizon of investment and the potential for additional investments. Investigations of partnering with others including qualified local partners should be considered.

8.4 Financial Analysis (Financial Model)

Financial analysis is the process of evaluating businesses, projects, budgets, and other finance-related transactions to determine their performance and suitability.

In some instants, financial analysis is used to analyze whether an entity is stable, financially healthy, or profitable enough to warrant a monetary investment. This could be an internal exercise requested by management to assess some scenarios per specific strategic objectives. It could also be in assessing possible investments through an acquisition or merger with another company.

Financial analysis with internal focus helps managers make future business decisions or review historical trends for past successes.

Financial analysis with external focus supports the enterprise in selecting among possible investment opportunities.

- Fundamental analysis uses ratios and financial statement data to determine the intrinsic value of security.
- Technical analysis assumes a security's value is already determined by its price, and it focuses instead on trends in value over time.

As an example, we describe the financial analysis with the associated financial model in the case of a "project financing" (noted in subsection 8.2.2). In that case the enterprise enters in a long-term agreement with an off-taker – for 100% of its production. This is through a product purchase agreement (PPA). The PPA stipulates the price and the quantities / specifications of the product and the deliveries. Other clauses relating to key obligations for both parties in the long relationship are included in the PPA. The project development team of the enterprise prepares a financial model that covers monthly outlays of expenses – during the construction and during the operation phase. During the construction phase, they factor the expenses for the EPC (Engineering, Procurement and Construction). The expenses also address expenses for the management of the development, advisors, owner engineer (as applicable), fees for permits etc. The payments received from equity and from the project loan are included in the model in accordance with the project financing agreement. The financial model depicts the payments and accrual of interest per the loan agreement (interest rate, term, etc.).

The financial model simulates the operational phase – monthly and annual expenses and revenue. Allowing for purchases, payments to staff and outsource services, insurance, and permits fees on the expense side. On the revenue side the model projects the sale of the production, and the inflows of funds from the sales. The model allows for planned and unplanned interruptions for maintenance. It allows for degradation, if any, that may result in decrease in revenue with time. Escalations in prices are considered.

With the projections of the expenses and revenue with time (yearly) the model projects the cash flow before tax and depreciation.

These come at the last stage of the model and then we get the projections for the rate of returns both on the capital and on the equity.

Generally, the financial model is set to enable varying input data to evaluate the impact on the returns depending on different scenarios.

The financial model is a very important tool in analyzing the project (technical analysis). In many instances, it is used between the sponsors of the project, the financial institution and the off-taker.

The above discussion applies to phases 1-3 of the project (subsection 5.2).

In the previous subsections we concentrated on financial aspects for the enterprise. It is the capital to establish the assets necessary for the start of the production. The nature of this is projection of the costs and the revenue as noted under this subsection 8.4. The calculation of project return is the main objective.

8.4.1 Accounting

When we look to the project implementation and execution phase (last phase subsection 5.2), we need to recognize the actual expenses. Before the commencement of the project operational phase budgets are prepared. The accounting staff gathers and records in the different ledgers all the payments. They keep a tab on all the expenses in different categories. These include external payments and internal (within the enterprise). The budget gives the appropriate guidance on the classifications of the expenses as needed for internal and external reporting.

The project lead accountant produces the periodical reports (usually monthly) to be reviewed by the project manager or department head. Some difficulties are faced in the delay of getting invoices and the actual dates for making the payments. These are challenges that require clear definitions in the reporting. The project manager or department head needs to have a definitive picture on the expenses to assess and report the standing versus the budget in a timely manner.

The skill of the accountants and their use of the appropriate software are important to get the desired objective. Very important is the timely reporting of the data – clearly and in a well-defined unit. If necessary, clarifying comments should be added.

The accountants make sure that all invoices are properly documented when presented to the financial institutions (for obtaining the loan withdrawals) per the procedures of the financing agreement. This is essential for timely collections and hence keeping the cash flow of the project on track.

When we get to the operation phase (phase 4 of the project – subsection 5.2), the manager or department head in the enterprise responsible for the plant is assisted by the accounting team to follow, the expenses and the revenue. The accountants compile expenses for all expense categories in details and in a timely manner. Through integrated business software, the integration of the accounting system with the entries of the various departments provides a dependable means with definite responsibilities. Through timely data entries in the integrated system, management gets reports both on revenue and expenses. Management can review the trends in a timely manner. Historical reports could be helpful for comparisons.

Besides the internal reporting, the accounting department also collects all the data that are to be included in the company financial quarterly, and annual reports for auditing and reporting to the tax authorities. For large enterprises with different branches and business lines complexities arise particularly when the enterprise operates across many countries with different tax systems.

8.5 Business Risks

Business risks encompass factors for a company (or organization) that would lower its profits or lead it to failure. Anything that threatens a company's ability to achieve its financial goals is considered a business risk. There are many factors that can add up to create business risk.

There is always some judgment in gauging the impact and the likelihood of encountering a specific risk and or group of risks. That is when the management skills come in – reviewing the results of the analyses and in many cases allowing for some margins/ requesting mitigation measures and or other action(s). Each enterprise has its own practice and puts in place procedures that identifies all the risks (internal to the enterprise), or external. Internal risks could cover items such as retention of personnel skilled professional, management (key personnel) and highly skilled technicians. The obsoleteness of the products may result in loss of market share. Aging/inadequate maintenance of equipment leads to production interruptions with loss of revenue. Violations to permits can be detrimental. Adherence to good practice as documented in policies and procedures minimizes risks.

External risks could come from interest rate change (enterprise can consider means to safeguard this or partly), climatic events such as storms/flooding etc. may result in business interruptions. Insurance coverage is a means of dealing of some the risks, of course at a cost. Change in law could be a major risk. Changes of the customers attitudes as well as competitor's actions in the market represent threats that could cause deterioration in the enterprise profitability.

8.6 Business Insurance

During the construction phase of the project, it is necessary to have in place "all risk policy" to cover personnel, the construction equipment at the site and the permanent equipment. The insurance for transport should be covered.

Appropriate knowledge within the enterprise to deal with the details in negotiating the insurance policies is essential. Engaging insurance consultants could be prudent. The goal is to reach the optimum insurance coverage to avoid excessive premiums while securing adequate coverage.

Once commercial operation starts, the enterprise normally secures new insurance policies with appropriate coverage commensurate with the nature of the operation. Inadequate insurance against events such as fire, theft, injuries of personnel, and in some instance business interruptions could have large impact to the enterprise returns.

It is important that the operations staff always stick to good practice to avoid/minimize events that lead to insurance company gets to handle such events. This is in the best interest of the enterprise.

8.7 Research and Development (R&D)

Research and development (**R&D**), known in Europe as **research and technological development** (**RTD**), is the set of innovative activities undertaken by corporations or governments in developing new products or services and improving existing ones. Research and development constitutes the first stage of development of a potential production process or new service.

R&D activities differ from an enterprise to another. In some case R&D is centralized, and in other cases it is decentralized. The R&D department (or a branch of the R&D department "Unit") could be staffed by scientists/engineers, The unit has the responsibility and is tasked with directly developing new products. The unit could be staffed with industrial scientists and tasked with applied research in scientific or technological fields, which may facilitate future product development.

The scale of the expenditures on R&D varies in different industries, but R&D is a must for the future of the enterprise. For a company, in good times, it is prudent to invest in what would fit in the company's future. The R&D staff should be acquainted with the problems in the production and in many instances get ideas from the production staff. Also, the R&D staff should follow research outside the enterprise and be cognizant with new inventions, products, material, and techniques for possible use.

R&D differs from the vast majority of corporate activities in that it is not intended to yield immediate profit, and generally carries greater risk and an uncertain return on investment. However, R&D is crucial for acquiring larger shares of the market and avoid stagnation that leads to negative results for the company.

8.8 Development of a 50 MW PV Solar Plant – (Example)

Following the discussions in subsection 8.4, a hypothetical case is chosen to demonstrate the development of a 50 MW Photo Voltaic PV Power Plant.

An International company X that has had extensive experience in development and ownership of renewable energy plants (both PV Solar and wind), it owns about 2 GW of – about 50 % PV and 50% wind. Company X has under construction about 900 MW. Its portfolio extends over 40 countries in the US, Europe, South America, and Africa. It's branch in Africa received an announcement for invitation to take part in pre-qualification leading to bidding for the development of the subject project. The review of the invitation for the PQ (Pre-Qualification) by the Africa team suggested that the company meets the required criteria – technical and financial. The Africa team asked for and got the concurrence of corporate to submit PQ for the project. As Company X did not have any previous presence in the project's country, its Africa branch had to get corporate permission. The received corporate approval was contingent on further review when the RFP for the project was to be received, provided company X made it through the PQ process.

The Africa team submitted the PQ for company X consistent with the requirements of the invitation to the PQ.

Company X received an official notification that it was among 9 companies that were qualified. It was also invited to purchase the RFP for the project for USD 1000. The due date for the submittal of the offer for this project was 3 months from the RFP date issuance. The Africa team proceeded to buy the RFP and immediately started the review of the RFP and per company procedures. The Africa team submitted the RFP summary covering the following among various points for a "bid / no bid decision", the following key points:

Summary of the RFP for Bid decision

- Plant Location
- Main scope – including connection to the Grid
- Capacity of the plant
- Site visit date

- **Key dates** – bid submittal date, announcement of winning bid, date of financial close, and Commercial Operation Date "COD"
- Term of the PPA (power purchase agreement) from COD
- Key Technical Data, e.g., panel specification, type of inverters, tracking, soil investigation report and grid specifications (including the grid code), any restrictions on auxiliary equipment

Commercial Conditions

Bid Bond
Performance Bond – for EPC and O&M periods
Liquidated Damages

- o Performance
- o Delay
- o Limits of L/Ds

Information about the Owner – Off-taker

- § Financial strength
- § Regulatory Body in Country
- § Political considerations – stability
- § Payments – currency / coverability
- § Participation of International Institutions (e.g.," International Financial Corporation" in preparation of the RFP)

Upon the approval of the executive management to take part in the bid, the bid preparation was kicked off.

A Proposal Manager "PM" for the development was appointed, and a bid team was mobilized. A technical manager that oversaw the preparation of the EPC and the O&M offers upon which the bid was to be based. In addition, a technical manager "TM", financial manager "FM" and legal manager "LM" were appointed. Under the PM oversight, the three TM, FM, and LM worked a bid preparation plan delineating the information including the timing of the exchange of the information to be provided from one group to the other.

Company X has two divisions (companies) that prepare contracts for the EPC and O&M. These two companies mostly execute these contracts for Company X. However, occasionally a third party is deployed.

TECHNICAL

The technical group prepared a tentative layout for the plant based on the information in the RFP. The group also prepared two **RFQs for EPC and O&M, respectively** that were transmitted to the EPC and O&M divisions of company X.

On the EPC and O&M side, from the divisions of EPC and O&M, two proposal managers were nominated, EPC PM and O&M PM. The two led the activities for the preparation of the EPC and O&M offers of the EPC and O&M divisions to be submitted to the PM / TM (for the development).

Under the EPC PM of the EPC division, the activities for the EPC offer progressed. First the design bases documents for the civil, mechanical (structural) and electrical disciplines were prepared in compliance with the RFP technical content. For unclear points, lists of questions were sent to the PM who in turn transmitted these questions to the client per the RFP procedures for requests of clarifications. The client per the RFP circulated answers to the requests of the clarifications from all bidders.

The EPC engineering team completed the technical contents of RFQs (request for quotation) for supplies (equipment, bulks) for the PV plant and the interconnection OHL (Overhead line) to grid connection point. and for the construction – including the Bill of quantities (BOQs). The procurement department in the EPC division issued RFQs to suppliers and subcontractors, delineating in addition to the technical requirements the commercial terms.

Main Supply packages:

- PV Panels
- Mounting Structures
- Central Inverters / Transformers stations
- CCTVs
- DC Cables & Accessories
- Combiner Boxes
- AC Cables
- Fence
- SCADA
- Weather station
- OHL

The technical team discussed with the procurement department the approach for the construction, it was decided that the EPC would be executed under the umbrella of the EPC division of Company

X, hiring local construction companies to act under the EPC contractor – from Company X – this approach was used successfully in many projects before.

Upon receiving the supply offers, the engineering disciplines completed the technical evaluations of the technical offers for the supplies – (equipment & bulks), 3 civil packages, 4 mechanical packages and 7 electrical. The number of offers for the packages averaged three responses from three different suppliers. The technical contents were reviewed and the offers that met the technical requirements were passed to the procurement department to complete the commercial evaluations for the supplies of equipment and bulks.

The procurement department coordinated with the EPC MP the procurement / delivery schedule of the lead items and that was integrated in the project EPC project that was prepared by the planner of the EPC.

In parallel, the final list of the selected suppliers was reviewed and approved by the PM of the EPC.

The construction offers were evaluated by the PM of the EPC and three local subcontractors were selected – civil, mechanical, and electrical.

The EPC team conducted a review of the selected offers and made final adjustments optimizing the schedule, and factoring prices between close contenders on both the supplies and the construction This led to final selection of the suppliers and construction subcontractors. The EPC coordinated further with the selected suppliers and construction contractors to get additional information to be included in the EPC offer to go to the development team.

The EPC offer was compiled addressing all the technical, managerial (execution; organizational, quality, safety- health, environmental) with details for the EPC schedule – milestones, and pricing, including milestone payments schedule.

The EPC offers acknowledged acceptance of the commercial terms included in the RFP of the EPC except for three items to which the EPC team chose to keep open for discussions.

The TM upon receiving the EPC offer from the EPC division reviewed the contents together with his group and prepared the following sets of documents:

i) Technical Input data to go to the Financial Modeler – please refer to financial discussions below
ii) Technical contents in the offer to be transmitted to the Client – Final review by the PM
iii) Technical & Commercial contents to be included in the EPC contract (between the development entity of company X and the EPC division of company X) – Final review by the PM and LM.

iv) The plant monthly electricity output for years 1 through 25 to be passed to the O&M team below. Allowing for the degradation.

v) The TM with his group prepared the technical contents for the offer to be submitted to the client based on the finalized contents of the EPC and O&M contracts.

Under the O&M PM (of the O&M division) the O&M offer was prepared in reply to the RFP received from the TM of the development. To a large extent, the preparation of the O&M offer follows similar steps to those undertaken by the EPC team described above.

The O&M built their plans for years 1 -25 of the plant following the completion of the EPC and successful completion of the project acceptance i.e. reaching commercial operation date "COD".

The O&M team submitted the following documents to the TM (development)

a) Mobilization plans (organization, manpower and related pricing)

b) Annual projections (manpower, consumables, spare parts, service contracts) - monthly cash flow for expenditures. The projections include replacement of equipment/ and or spare parts) per the recommendations of the suppliers for year 1 through year 25.

c) The projections of the plant monthly output per the engineering outputs from the EPC division (item iv above) are reviewed and accepted. These would be included in the O&M contract.

d) The contract for the O&M was prepared and submitted to the PM/TM of development at company X. The contract specified quarterly payments, covering manpower staff, spare parts, consumables, outside services including security.

FINANCIAL

The FM with his group followed these tasks:

1) Setting up of the financial model for the project.

The model input included:

Schedule

Expenses – Capital; EPC payments (per the EPC contract item iii above), development expenses including studies, permits, and consents.

- Operational Expenses; (per the O&M contract item d above), land lease plus insurance, SPV operations costs.

Production – the data received from c) above for the electricity production per month.

Incorporating the following factors in accordance with the PPA received from the client:

Payment terms / collection

Exchange Rate of Currency

Inflation

Loan Terms: Interest Rate, Term including the terms for the construction period and the term of the loan.

Monthly projections of the expenses.

Projections of the revenue based on TARIFF to be included in the offer to the client.

Company profit taxes.

Factor gives the profit margin expected from the project that is incorporated in the TARIFF.

2) Preparing the financial solicitation of financing that was submitted to candidate financial institutions to finance the debt – target was 70 /30 Debt / equity financing.
3) Interfacing with the financial institutions and negotiating the terms for the financing package for the project, leading to receiving a commitment letter for project financing.
4) Completing the commercial inputs to the development offer for submission to the client.

LEGAL

The LM with her group undertook the following tasks

I. Review of the RFP from the client including the PPA and the grid interconnection agreement.
II. Review of the PPA annexes that were completed by the TM, LM.
III. Participation with FM in the negotiations of the financial loan agreement.
IV. Setting up the SPV (Special Purpose Vehicle) company for the project.

BID ENVELOPES

Technical – Plant & OHL descriptions – equipment specs

EPC plan

Tests for plant acceptance

Plant dispatch

Commercial – including bid bond
Tariff
Interfacing with the off-taker – invoicing, collection

The PM, TM, FM and LM convened a final review meeting in which all aspects of the preparation of the bid were discussed and the PM gave instructions to resolve and clarify specific points that were included in a list.

The PM prepared a brief for the management committee meeting to get the authorization for the submittal of the bid in the name of company X to the client.

The bid of company X was submitted to the client and the opening bid was attended by the PM on behalf of company X. Seven bidders submitted bids including company X.

Fortunately, the TARIFF of company X was the lowest.

The client reviewed the bids received and during this review time Company X received a list of questions which the PM coordinated the preparation of the responses. Then he transmitted the responses to the client.

The client issued the results of its evaluations of the bids and announced that company X is the nominee to undertake the project. Company X was invited to a precontract meeting in which all points to sign the contract were discussed and agreed to.

The incorporation of the SPV was completed and the principals of the SPV were nominated and started taking charge of the execution of the project.

The development team finalized all documentations relating to the project including the EPC and the O&M, the financing agreement, the insurance contract, and took care of advance payments and issuance of bonds. All contracts were signed satisfying the order of precedence. With the conclusion of the signing of the project agreement by the SPV principals the development team turned over all the project documents to the SPV and the development was completed.

8.9 References & Bibliography

1) **J. Fred Weston, Thomas E. Copeland**, *Managerial Finance*, Eighth Edition, The Dryden Press, CBS Publishing Japan Ltd., ISBN 0-03-064041-5, ISBN 0-03-910728-0
2) **H.H.Safwat** ,"Independent Power Producers", Encyclopedia of Energy, Engineering and Technology, Vo. 1-3, Editor Caphart Barney.
3) **H.H. Safwat,** "Engineering, Procurement and Construction of PV Parks", , Solar Energy, Cairo, Egypt, October 2015.
4) https://www.investopedia.com/economy-4689801

PART IV

COMBINED OUTPUT FROM KNOWLEDGE & ENERGY & RELATION TO ECONOMICS

Under Part IV, in section 9, we build on the quantitative K&E model of section 5, and the information in parts I, II and III introducing the quantitative K&E model. This is the key contribution of this book. In section 10, conclusions and recommendations are included with specific reference to contents in the ***Reference.***

Section Nine: IT IS ALL ABOUT JOULES

The author together with Professor Dr. Oweiss started their effort to examine the similarities between economics and thermodynamics in 1988. In 2002 they managed to publish in the **Reference** (please refer to subsection 9.1,) the results of their research as noted in section 1. In the **Reference**, the authors invited scholars to examine the topic further in search of both better understanding and possibly develop a means/ methodology that can be applied in economic studies.

The models proposed in the Reference were preliminary and may have lacked some rigor.

Both the author and Professor Dr. Oweiss recognized this shortcoming. This is what prompted the two authors to continue their search for a more practical means. Unfortunately, the author really missed the insight of Professor Dr. Oweiss as he became not well in recent years. As noted in section 1, this book represents the second attempt to add to what was sought since the original idea first identified by the author and professor Dr. Oweiss, about 30 years ago.

The information in sections 2, 3 and 4 have led to the formulation of the K&E interplay model that is presented in section 5. In section 6, the K&E interplay model was illustrated in several situations. Under the K&E interplay model, and noting the overall objective is attain highest efficiency in obtaining output from inputs. This translates to minimize the knowledge and the energy combination. The premise of representing the effect of the knowledge by an equivalent energy emerged. This is the foundation of the discussions in this section.

Based on the information presented in the previous sections, the following subsections have been formulated. In subsection 9.1, some key points from sections 1-6 that form bases for the quantitative K&E are noted. The quantitative K&E model is presented in subsection 9.3, following subsection 9.2 relating knowledge to labor information (presented in subsection 7.4). Relating to information in sections 7 & 8, Subsection 9.4 covers additional factors from the cost perspective and in subsection 9.5 additional considerations for profit and taxes are addressed.

9.1 Bases for the Quantitative K&E Model

In discussions with Professor I. Oweiss when preparing the Reference in 2000, the following conclusion was reached:

"Going back to the start of bartering, when one person was bartering his cotton with another person that had wheat, what they were exchanging were not the products cotton versus wheat. It was the knowledge of planting cotton versus the knowledge of planting wheat. This is the essence of knowledge being behind the products and services that have been and continued to be developed and exchanged (traded) for centuries".

In recent years the theme we are now living in the knowledge age has been repeated in many circles. The reality is that since the early days of known history, knowledge was there and has played a vast part of human development. What we can say is "now we have better recognition of the role of knowledge in our lives". Certainly, this book underlines the relevance of knowledge practically in all activities".

From sections 2 and 3 we see that knowledge and energy have large similarities.

a. Source (energy & knowledge)

The source of energy is from nature (e.g. fossil fuel, or solar irradiation and wind). The source of knowledge is originating from the brains of humans. Both energy and knowledge are not substances (material) and only their effects can be sensed. In the case of fossil fuels, it took nature a very long time to form the fuels that are extracted from the ground or from under the sea. For knowledge, over centuries knowledge was gathered in various forms and was characterized in different topics. The body of knowledge we have today is a part of a continuum. The process of knowledge build up is quite dynamic and more information joins the body of knowledge all the time. The evolution of education in schools and colleges has led to larger percentage of the population of the world now to reach much higher levels. This automatically contributes to the spread of knowledge. In parallel, thanks to technological methods in mining and oil/gas extraction an abundant amount of fossil fuels is now available. In the last twenty years the use of renewable energy from the sun and the wind made significant strides. The renewable resources are dispersed and intermittent. The extraction of fossil fuels and its preparation to suit the uses in power plants and transport means has been made possible by technologies derived from knowledge.

b. Low Grade Form (knowledge & energy)

Both knowledge and energy have low-grade forms. That is very clear for energy. When we start with internal energy in the fossil fuel (e.g. an amount of coal or oil) the content has internal chemical energy. That energy, when liberated by burning, results in heat. Please refer to the definition of "Heat" in subsection 2.1. As demonstrated in section 2, the heat engines/thermodynamic cycles are used to convert heat low-grade energy (since it has limited uses) to mechanical/electrical energy "work" that can be utilized for many purposes. A similar logic follows when we address the dispersed energy form in solar energy isolation reaching a surface or wind blowing, these energies are in forms that are difficult to put into use. Through photo voltaic (PV) and wind turbines, we transform these low-grade energies to electricity.

A parallel can be drawn for knowledge that is coming from "the thinking of thousand/millions of humans". The analogous of thinking in the heads of the human is chemical energy in the fuel. Unless the thinking of humans is funneled in appropriate ways of building knowledge, the thinking would remain scattered. The scattered thinking would be like the case of fossil fuel staying in the ground. The scattered thinking is low-grade knowledge.

c. Brain & Human Body

In section 3, we have seen that the brain performs many functions. One of them was identified and discussed; the controls the food energy as it converted in the other organs of the body. The converted energy now becomes "work", the term defined in section 2. Neurons in the brain produce electrical signals. This we can refer to as high-grade energy that organs use in the performance of various body functions; "premium energy". As was noted, the energy of the food is in Joules, and these are low-grade energy. The human body systems transform that energy to high grade energy or "premium energy". The premium energy is consumed by various body systems. As noted in section 3, the brain uses part of this premium energy.

From the comparison of the brain and the computer in subsection 4.1 we recognize that brain memory is storing "knowledge". The brain functioning together with the rest of the body systems is a demonstration of knowledge-energy combining in the execution of the various activities the human undertakes.

d. Transformation from Low Grade to Premium

From subsection 2.1, through the devised thermodynamic cycles, "heat" low-grade energy is converted to "work", then "electricity" or premium energy. According to the second law of thermodynamics, the higher the temperature of the added heat, the better the efficiency of the thermodynamic cycle. In subsection 2.2 the application of the thermodynamic principles in power plants (generation) was demonstrated. As was noted in section 2, the premium energy is produced as work/electricity to transport it to various users. Also, noted in section 2, "net-work" from the power plant is equal to turbine generator output minus the power consumed internally in the plant (by the auxiliary equipment). Behind all power plant applications are the technologies that are produced from knowledge. Reaching high temperatures and maintaining these high temperatures are products of innovations from engineers and metallurgists. Through the technologies with new innovations higher thermal efficiencies have been attained, (useful premium energy "work" relative to added heat (low-grade energy). We have learned from the second law of thermodynamics some heat is rejected as inherent energy waste.

The same parallel transformation in the human body is highlighted under c., we note the body produces waste.

e. Knowledge and Innovations

Subsection 5.5 presented knowledge. We identified the terms falling under it: information, data. We further discussed "technology" as a manifestation of knowledge to achieve desired results. In subsection 5.6 we discussed "innovation," the schemes that are introduced through knowledge to improve or reach new uses. Examples of technologies correspond to what we see in the adoption / adaptations of thermal cycles in power plants. Innovations can be considered part of knowledge, or it is closely related to knowledge.

Through innovations we can get to higher levels of knowledge that are key to more enabling technologies.

f. Knowledge & Energy Interplay

In subsection 5.1 the K&E Interplay model was introduced. Knowledge adaptations to attain technologies has been discussed, to put technologies to use always involves energy. The role of technologies in advancing power plants was highlighted.

g. Efficiency

In section 2, we discussed the term thermal efficiency (and the term performance) – the relative ratio of the "net-work" to the added heat. Thermal efficiency is a measure of achieving the objective of getting premium energy from a low-grade energy. It is the ratio between "desired output to the input". Note, the inherent un-used fraction "heat rejection" or waste in the case of heat engine and the human body, respectively.

We shall build on this as we introduce a generalized definition for efficiency to be adopted when dealing with products and services through combining "knowledge & energy".

h. Representation of Knowledge and Energy in Joules

Unlike energy that has a well-established unit "Joule" which has been in use and serves as a basis of all the energy quantification, we lack a quantification unit for knowledge.

As the proposed under the knowledge-energy interplay model, the two components are present in all activities. In the K&E model applications presented in section 6, the two components are present in each action / task. In some situations, knowledge is dominant, in other cases, energy takes a bigger role. K&E are present together all the time. In any enterprise, the measure of the output to the input is always sought – the "efficiency".

With the close similarities between knowledge and energy as summarized in items a through f above, the author proposes to deal **with knowledge and energy with the commonly used unit of energy "Joules".** The author realizes that this proposition is quite a drastic shift in many respects.

Additional ideas behind this proposition will be clear from items i and j that follow:

i. Minimizing sum of Inputs for Products and Services as outputs of Enterprises

Always in any enterprise, getting the product or the service with the optimized sum of the knowledge and energy is on the mind or the enterprise principals. Under the proposition in h above, this could be stated that the aim is to **use the minimum sum of inputs (combined) from technology and energy in Joules to produce a product or a service for a specific value of Joules.**

From sections 5 and 6, the applications of the K&E model to variety of enterprises demonstrate the simultaneous contributions of the K&E in all activities.

j. Knowledge and Labor

From section 7 "Macroeconomics" we saw how labor is related to knowledge. In subsection 7.4 we see the higher wages commanded by occupants in management, financial and information services give indications to the value-added of their contributions.

From subsection 7.15, we see the contributions of other industries to any specific industry. Please refer to Tables 7.15.2.1 and 7.15.3.1. These are in line with the schematic presentations of the K&E model for manufacturing and services in section 6. Thus, we have in the enterprise internal "i" knowledge contribution and direct energy contribution. These two internal contributions are the sums of each throughout the activities undertaken for a specific product or service. Under this terminology, all the knowledge & energy contributions/consumptions from different participating internal departments/personnel are added.[1] For externally obtained intermediate products/services the same is to be assessed for each intermediate input the enterprise acquires that end up in the completion of the output, termed external "e".

---

1 Generally in many industries "direct" is for charges coming from the production departments, while indirect comes from central functions e.g. management, support departments etc.

Subsection 9.2 builds on the above to arrive at estimates for electricity amounts corresponding to knowledge contribution in an activity.

9.2 Contributions of Knowledge and Energy in an Activity

Generally, an activity is completed in a period; during this activity one can ascertain the amount of the electrical energy that was consumed. This is an amount of E_a kJ (energy for activity a).

Turning to knowledge, it is more complex, let us assume one person involved in the execution of this activity spent n hours. What he/she was delivering during this n hours is a result of his accumulation of knowledge for years. Now he/she is, through the person's brain, using the knowledge in the delivery for the activity. What the person is delivering in the n hours amount to K_a kJ. What the person is delivering is dependent of his knowledge level. As we saw in section 3, the brain stores info. It also controls the body systems/organs as these systems perform their functions. That is energy.

>>>>>>> This is the reason the author is proposing dealing with knowledge in joules.

This leads to the presentation of an activity of duration n hours, we are dealing with the sum of

$$K_a \text{ kJ} \ \& \ E_a \text{ kJ}.$$

For K_a kJ the measure is more complex, particularly as the above approach is being newly introduced. If historical data are available, then K_a and E_a can be obtained / calculated from the data.

The author turned to the wages that were presented in subsection 7.4.

While knowledge has a lot of similarities with energy as noted in subsection 9.1, yet there is a difference when we take a deeper look in item e of subsection 9.1. With knowledge accumulation and innovation, there will be different deliveries for personnel. This suggests a "ranking"; let us refer to this ranking by knowledge grade for personnel. This really comes down to the workforce occupants. Specifically, the capability of a person manifested in his/her knowledge to perform in the position he/she takes in an enterprise or in the government. In absence of a grades scale for knowledge (together with innovation), the closest scale the author can propose is to use the labor wages in subsection 7.4.3 (please refer to the wages of table 7.4.3.1). In table 9.2.1, the rows of table 7.4.3.1 were grouped to replace 21 rows by six rows (for simplicity in demonstration examples that follow) and the four columns in Table 9.2.1 provide the results of this grouping.

Then assuming 1900 hours/year and an electricity price of 0.125 $/kWh the values in the last two columns were calculated, the last column is the equivalent Knowledge kWh for an hour worked by the person in a particular group.

Please note that the origin of the equivalent knowledge EqtK as proposed by the author below is energy of the human body. The energy is not really exerted by the person during the hour, but it is an equivalent energy that signifies energies associated with the knowledge that person accumulated in his/her repository in the brain and is pulling out during the execution of the activity/ hour.

Table 9.2.1 Demonstrative example of the Proposed Grade scale for knowledge

% of Group to Total Occupants	Group Wage $/year	GRADE	Ratio to Wage of Highest Group	$/h	EqtK Equivalent kWh
9.5%	100,260.73	Grade 1	100%	52.77	422.15
2.5%	81,564.87	Grade 2	81%	42.93	343.43
12.0%	68,937.20	Grade 3	69%	36.28	290.26
17.8%	48,638.12	Grade 4	49%	25.60	204.79
28.8%	39,848.46	Grade 5	40%	20.97	167.78
29.4%	27,354.24	Grade 6	27%	14.40	115.18

Examining Table 9.2.1 we can say Grade 1 is more valuable than the succeeding grades. This translates to larger Equivalent kWh "EqtK" (Equivalent Energy to Knowledge). It is noted that we could find extra ordinary knowledge that would have much higher/superior grade that are not included in the demonstration Table 9.2.1, e.g. in arts, or professional sports or top surgeons. We are talking of exceptional knowledge of individuals that can master the delivery of this knowledge at exceptional levels.

An analogous weak characterization for energy is not quite apparent, but it is not important as in an enterprise we primarily deal with electricity. The focus here is about electricity when it reaches enterprises. However, after a closer look, we find the electricity companies, TSO and DSO, introduce different tariffs for daytime use versus night-time. Further TSO offers different tariff to enterprises that take their electricity at high or extra high voltage. Accordingly, below in Table 9.2.2 we may use factors for electrical energy depending on the characteristics of the use, if this need to be considered.

Table 9.2.2 Energy Characterization Factor

Usage Characterization	EUF
U1	1
U2	1.05
U3	1.1
U4	0.95

The values in Table 9.2.2 for the Energy Characterization Factor "EUF" are assumed. They are to depict normal usage "U1", then use during other times, daily peak, summer daily peak and night use, respectively for U2, U3 and U4 respectively.

9.3 Quantitative K&E Model

From the K&E Interplay model introduced in subsection 5.1 and based on the arguments enumerated in subsection 9.1 together with the rationale proposed under subsection 9.2 above, we can write the following equations

$$\textbf{INPUT} = I = K + E \qquad \text{Eq. 9.1}$$

Where

K is the knowledge Input kJ

E is the Energy Input kJ

Based on the discussions of items i and j in subsection 9.1, we are dealing with two types of value-added for a product or service an internal and external in each of KI and EI,

These are

ki = internal knowledge kJ

ke = external knowledge kJ

and

ei = internal energy kJ

ee = external energy kJ

For each product we need to make the summations for the total number of internal activities N, and depending on the number of the external inputs say Y each with its own M_Y activities

$$K = \Sigma_N \, ki + \Sigma_{yM} \, ke \qquad kJ \qquad \text{Eq. 9.2}$$

&

$$E = \Sigma_N \, ei + \Sigma_{yM} \, ee \qquad kJ \qquad \text{Eq. 9.3}$$

It is important to note that when we add the energy amounts – we must have uniform type of energy – e.g., for all the summed energies it is electricity. We should not mix heat and electricity. This is the essence of the different levels (grades) of energy that had been frequently mentioned in this book.

The same rule applies for knowledge, when we deal with different levels of energy or knowledge we should introduce the appropriate conversions for the level's. Unlike the case of converting low energy to premium energy in heat engines, for an enterprise we are mostly dealing with premium energy with respect to both knowledge and energy; hence the objective reduces to optimizing the sum of the INPUT i.e., the sum of K + E. Kindly, refer to Figure 1.3.1.

Please note that we have always to contend with inherent losses in processes; these we try to minimize, but we can't avoid altogether. In the K&E model we are dealing with the Inputs--not the inputs after taking out the losses. By taking the necessary steps to minimize wastes we are indirectly getting low inputs.

Now we are to look for ways to reduce the combinations that come from the right sides of Eq. 9.2 and Eq. 9.3 as we plug the values of K and E in Eq. 9.1

To attain lower sum for inputs, outsourcing in some cases may be an appropriate means; a new/ innovative solution may be another way.

Referring to subsection 6.2 where the importance of historical data was underlined, to execute the above procedure, the values for the various activities need to be obtained for both the internal and the external records. For the external there could be some difficulty obtaining the details from sub-suppliers / subcontractors, but they could be willing to share the required data.

For an enterprise that has been producing a particular product, the enterprise has from historical data the data for the knowledge and energy. Granted in the first application of the K&E model, the values for k will be more difficult as being estimated for the first time. Let us say the enterprise, now is considering use of a newer technology for some activities, then the exercise reduces to evaluating the sum of the new K and E with the newer technology versus the current sum of K and E.

The investment considerations discussed in section 8 would help in making the final decision to go for the adoption of the newer technology or not.

To demonstrate the above logic, three examples from three different industries are analyzed below making assumptions with many simplifications, as the objective is merely to demonstrate the approach. The author had to turn to the sources of data in the public domain, adding to what was done in subsection 9.2.

For the author, in absence of historical data, to demonstrate the application for manufacturing we refer back to the case of pipe spool fabrication subsection 6.1.1. For services we consider the one example of the medical family doctor of subsections 6.1.2. The following application is simplified by considering only a limited number of activities. This suffices for the purpose of the demonstration.

Referring to the industrial classifications referred to in subsection 7.15.1, the first example falls under classification 322 and the second under classification 622.

The corresponding external/internal industries matrix (like Table 7.15.2.2) for these two industries can be obtained, under classifications 332 and 622, respectively. Please refer to Table 9.3.1 below

Table 9.3.1 Matrix of Industries for Fabricated Metal and Hospital – Industry Classifications 332 and 622

	IOCode	**332**	**622**
Industries/ Industries	**Name**	**Fabricated metal products**	**Hospitals**
111CA	Farms	**0.0186780**	0.0000001
113FF	Forestry, fishing, and related activities	0.0053227	0.0000001
211	Oil and gas extraction	**0.0363401**	0.0000001
212	Mining, except oil and gas	0.0295691	0.0000001
213	Support activities for mining	0.0177229	0.0000001
22	Utilities	0.0092765	0.0000001
23	Construction	**0.0677003**	0.0000001
321	Wood products	**0.0433282**	0.0000001
327	Nonmetallic mineral products	**0.0384641**	0.0000001
331	Primary metals	**0.0674761**	0.0000006
332	Fabricated metal products	1.1369962	0.0000002
333	Machinery	**0.1119231**	0.0000001

334	**Computer and electronic products**	**0.0158569**	0.0000000
335	**Electrical equipment, appliances, and components**	0.0963702	0.0000001
3361MV	**Motor vehicles, bodies and trailers, and parts**	**0.1337794**	0.0000001
3364OT	**Other transportation equipment**	**0.0691156**	0.0000001
337	**Furniture and related products**	**0.0700902**	0.0000001
339	**Miscellaneous manufacturing**	**0.0492637**	0.0000001
311FT	**Food and beverage and tobacco products**	**0.0376038**	0.0000001
313TT	**Textile mills and textile product mills**	**0.0297823**	0.0000001
315AL	**Apparel and leather and allied products**	**0.0225707**	0.0000001
322	**Paper products**	**0.0559517**	0.0000002
323	**Printing and related support activities**	**0.0357345**	0.0000001
324	**Petroleum and coal products**	**0.0243612**	0.0000001
325	**Chemical products**	**0.0294938**	0.0000001
326	**Plastics and rubber products**	**0.0515052**	0.0000001
42	**Wholesale trade**	0.0067979	0.0000001
441	**Motor vehicle and parts dealers**	0.0093299	0.0000001
445	**Food and beverage stores**	0.0067796	0.0000001
452	**General merchandise stores**	0.0051653	0.0000001
4A0	**Other retail**	0.0086784	0.0000001
481	**Air transportation**	0.0063587	0.0000003
482	**Rail transportation**	0.0166234	0.0000001
483	**Water transportation**	0.0472491	0.0000003
484	**Truck transportation**	0.0132159	0.0000001
485	**Transit and ground passenger transportation**	**0.0114747**	0.0000001
486	**Pipeline transportation**	**0.0186162**	0.0000000
487OS	**Other transportation and support activities**	0.0207803	0.0000003
493	**Warehousing and storage**	**0.0140574**	0.0000001
511	**Publishing industries, except internet (includes software)**	0.0058800	0.0000000

512	Motion picture and sound recording industries	0.0044054	0.0000002
513	Broadcasting and telecommunications	**0.0234154**	0.0000001
514	Data processing, internet publishing, and other information services	**0.0116450**	0.0000001
521CI	Federal Reserve banks, credit intermediation, and related activities	0.0033683	0.0000001
523	Securities, commodity contracts, and investments	0.0046887	0.0000002
524	Insurance carriers and related activities	0.0028042	0.0000000
525	Funds, trusts, and other financial vehicles	0.0043520	0.0000002
HS	Housing	0.0027319	0.0000000
ORE	Other real estate	**0.0129505**	0.0000001
532RL	Rental and leasing services and lessors of intangible assets	0.0071408	0.0000001
5411	Legal services	0.0029301	0.0000000
5415	Computer systems design and related services	0.0044324	0.0000000
5412OP	Miscellaneous professional, scientific, and technical services	0.0080435	0.0000002
55	Management of companies and enterprises	0.0062595	0.0000001
561	Administrative and support services	0.0097081	0.0000003
562	Waste management and remediation services	**0.0344635**	0.0000002
61	Educational services	0.0085744	0.0000001
621	Ambulatory health care services	0.0053263	0.0000345
622	Hospitals	0.0076067	1.0028171
623	Nursing and residential care facilities	0.0066677	0.0000006
624	Social assistance	**0.0148454**	0.0000000
711AS	Performing arts, spectator sports, museums, and related activities	0.0037007	0.0000014

713	Amusements, gambling, and recreation industries	0.0095463	0.0000001
721	Accommodation	0.0088509	0.0000001
722	Food services and drinking places	**0.0107232**	0.0000001
81	Other services, except government	0.0134527	0.0000003
GFGD	Federal general government (defense)	0.0184881	0.0000001
GFGN	Federal general government (nondefense)	0.0053990	0.0000034
GFE	Federal government enterprises	**0.0153016**	0.0000001
GSLG	State and local general government	0.0094260	0.0000092
GSLE	State and local government enterprises	**0.0310748**	0.0000001

Examining Table 9.3.1 reveals that for classification 622 there is very little dependence on other industries. This means in the simplified application here we are to neglect the external items.

For the Fabricated metals, for the demonstrative example, the values in column 322 in Table 6.3.1 were ordered in a descending order, then three industries were taken as external industries for the demonstrative example of the fabricated pipe spools. Please refer to table 9.3.2.

Table 9.3.2 Three selected external industries to be considered in the pipe spool fabrication:

	IOCode	332	
Industries/ Industries	Name	Fabricated metal products	To Take
332	Fabricated metal products	1.1369962	
3361MV	Motor vehicles, bodies and trailers, and parts	**0.1337794**	Take
333	Machinery	**0.1119231**	
335	Electrical equipment, appliances, and components	0.0963702	
337	Furniture and related products	**0.0700902**	
3364OT	Other transportation equipment	**0.0691156**	

23	Construction	0.0677003	Take
331	Primary metals	0.0674761	Take
322	Paper products	0.0559517	

9.3.1 Example #1 Pipe Spool Fabrication

Under the following assumptions an estimate is made for the shop floor work as tabulated in Tables 9.3.1.1, 9.3.1.2 for the internal activities.

The company has won a bid to supply the following three different types of pipe spools; for simplicity these are referred to 12", 10" and 6 ". The numbers of each of these spools is 10, 24 and 80, respectively.

The engineering department and the planning departments have completed the design and the manufacturing plans. Further, all the required material has been ordered and have been received to be released from the store in the shop.

Table 9.3.1.1 Estimates for the Knowledge and Energy consumptions for one 10" pipe spool (internal)

SHOP FLOOR	KNOWLEDGE					ENERGY			
Operation	Person	Grade	Duration h	Eq. kWh/h	kWh	Machine	Duration h	Power kW	kWh
Cutting	Mechanic	Grade 5	3	167.78	503.3	Heavy Duty Saw	3	50	150
	Helper	Grade 6	1	115.18	115.2				
Bending	Mechanic	Grade 5	4	167.78	671.1	Heavy Duty Bender	5	20	100
	Helper	Grade 6	1	115.18	115.2				
Welding	Mechanic	Grade 4	6	204.79	1,228.7	Welding Machine	4	20	80
	Helper	Grade 6	2	115.18	230.4				
Transport	Operator	Grade 5	4	167.78	671.1	Crane	3	40	120
	Helper	Grade 6	4	115.18	460.7				

QC	Inspector	Grade 5	4	167.78	671.1	Mea-suring Equip-ment	4	0.2	0.8
				SUM	,666.9			SUM	450.8

The estimates in Table 9.3.1.1 indicate the knowledge equivalent kWh is ten times that of the energy side for the 10" pipe spool.

Without repeating the same exercise for the two other sizes 12" and 6" a factor[1] is applied to the 10" results and the summary of the results are included in Table 9.3.4.

Table 9.3.1.2 Summary results for Shop Floor Activities - Internal

Spool Size	Factor	Knowledge kWh/ spool	Energy kWh/ spool	Combined kWh / Spool	Number of Spools	Knowledge kWh	Energy kWh	Combined kWh
12"	1.5	7,000.3	676.2	7,676.5	10	70,003.2	6,762.0	76,765.2
10"	1	4,666.9	450.8	5,117.7	24	112,005.1	10,819.2	122,824.3
6"	0.75	3,500.2	338.1	3,838.3	80	280,012.8	27,048.0	307,060.8
					TOTAL	**462,021.1**	**4,629.2**	**506,650.3**

The contents of Table 9.3.1.2 indicate the following:

$$ki = 462{,}021.1 \text{ kJ}$$

&

$$ei = 44{,}629.2 \text{ kJ}$$

Now we turn to estimating external knowledge and energy; this is challenging without having access to real data from suppliers, but for the demonstrative purposes, some simplifying assumptions are being made:

Referring to Table 9.3.2 corresponding to the three parameters marked with "Take", we use the values in the table as the external participation factors. However, judgment was made with respect to the second take, and the assumed value is entered in red. Please refer to Table 9.3.4.

Table 9.3.1.3 Estimate of External knowledge & energy for the pipe spool example

Industry Class	Industry	External[3] Participation	External % relative to Internal	Knowledge[5] % of Combined	Energy % of Combined	Combined kWh	Knowledge kWh	Energy kWh
				INTERNAL[2]				
332	Fabricated metal products	1.1369962		90%	10%	**506,650.3**	**462,021.1**	**44,629.2**
				EXTERNAL				
3361MV	Motor vehicles, bodies and trailers, and parts	0.1337794	11.8%	75%[5]	25%	59,612.7	44,709.5	14,903.2
331	Primary metals	1.5[4]	131.9%	20%	80%	668,406.3	133,681.3	534,725.1
322	Paper products	0.0559517	4.9%	40%	60%	24,932.3	9,972.9	14,959.4
					Sums	752,951.3	188,363.7	564,587.6

2 From Table 9.3.1.2

3 External participations are taken from Table9.3.2, except for the row of classification 331 – primary metals

4 Value in red is assumed

5 Values in blue are assumed

From Table 9.3.1.3 we get, the external kWh

$$ke = 188{,}363.7 \text{ kWh}$$

&

$$ee = 564{,}587.6 \text{ kWh}$$

Combining the results of internal and external knowledge and energy kWh for the simplified pipe spool example are summarized in Table 9.3.1.4.

Table 9.3.1.4 Summary of the Results for the Pipe spool Example

li = ki+ei kWh	ki kWh	ei kWh	ki/ li	ei/li
506,650.3	462,021.1	44,629.2	91%	9%
le = ke+ee kWh	ke kWh	ee kWh	ke/le	ee/le
752,951.3	188,363.7	564,587.6	25%	75%
l= li + le kWh	ki+ke kWh	ei+ee kWh	(ki+ke) /l	(ei+ee)/l
1,259,601.7	650,384.82	609,216.84	52%	48%

The top row is the internal knowledge and energy estimates. The grey rows are the external knowledge and energy. The bottom rows correspond to the sums of the internal and external knowledge and energy. The last columns give relative values to the inputs.

9.3.2 Example #2 Medical Family Doctor

Let us follow a typical working day for Dr. John SMITH at his clinic in suburb Nice of City in the State representative in the USA.

18 patients visited the clink and he spent on the average 20 minutes with each. The doctor prepared 18 visit reports for each of the patients he saw that day.

The nurse coordinated the entries of the patients to the doctor's office. She has completed the paperwork for each patient prior to entry to the doctor. She ordered six lab tests for three patients that the doctor requested. She also answered the phone for reservations of appointments in the following days. No cancelations occurred that day. She went through the mail and got the organized the reports prepared by the doctor for forwarding to the insurance. She collected the visits fees from the patients. She oversaw the helper that is responsible for the cleaning of the clinic and other miscellaneous duties.

Table 9.3.2.1 Projections for Example 2 Medical Family Doctor

	Knowledge				Energy			
		Duration h	Eq. kWh/h	kWh		Duration h	Power kW	kWh
Doctor	Grade 2	8	343.43	2,747.4				
					Computer	8	0.2	1.6
					Medical Equipment	2	0.25	0.5
Nurse	Grade 4	8	204.79	1,638.3				
					Computer	8	0.2	1.6
					Med. Equip.	2	0.4	0.8
Helper	Grade 6	8	115.18	921.4				
					Equipment	3	1	3.0
			SUM	5,307.2			SUM	7.5

Per Table 9.3.1 the external industries participation is negligible. However, the author feels when actual historical data for a real case would be available, there will be external effects to be factored.

For the time being, for this simplified example, the knowledge component is quite dominant compared to the energy, with a factor of almost 100 between the two. The two demonstrative examples in this subsection had several assumptions, hence the results could only be judged as indicative. In the context they are meant to represent the logic the author has followed in applying the suggested approach.

9.4 Innovations & Knowledge Application

In subsection 9.3 the author had to resort to workforce wages in the labor statistics. The author noted that historical data for wages at a particular enterprise if available would have been a better source. Yet there remain some challenges that were flagged in subsection 9.2 and 9.3--"the grading of knowledge".

Unlike energy in power generation where we had straight forward units, for fuel we could perform tests and quantify the heat value for any fossil fuel (measurements). For electricity, we have physical characteristics, the frequency for AC (alternating current) and the voltage level. We face a multitude of challenges as we try to address the grades of knowledge. The proposition in subsection 9.2 to use equivalent Joules to represent knowledge calls for addressing and incorporating the factors numerated below to reach the sought knowledge grades.

i) We start with the source of knowledge is in the brains of humans.

A. SCHOOL AND UNIVERSITY GRADES

ii) In schools and colleges grades are given by the teacher / professor as an indication of the student's comprehension of the course.

iii) Further, an average grade for the courses the student takes in a semester is always recorded.

iv) For the stages of schooling and college, an average still with underlining the grade in the last year before graduation.

v) The grades noted under ii) – iv) all refer to education – part of education.

B. TRAINING AND GRADES OF TRAINING PROGRAMS

vi) A person may take training courses to get certain skills or enhance his familiarity of these skills, e.g., computer skills, language, job related skill or in some cases training on complex equipment operation, or maintenance.

vii) For many professions, periodical training requiring certifications is a must, e.g. certified accountants, lawyers, dentists, and professional engineers.

viii) Many professional associations serve in this area, also in some cases branches of universities and colleges offer training programs. Some programs are available on the internet (virtual).

ix) Trainings in communications, verbal or written, and in making presentations are valuable.

C. ON THE JOB LEARNING

x) Without a doubt, learning on the job provides the most dependable knowledge that stays with a person. It is through the repetitive handling of the tasks, that various aspects of the knowledge sinks in. The brain collects and organizes that knowledge, and the brain calls back the stored knowledge whenever needed.

xi) Aids to the practicing person, could be in the form of manuals, videos, notes etc.

xii) In many situations the time the person spends in the same job, gives him more in-depth knowledge of the job.

D. GENERAL KNOWLEDGE

xiii) There is beneficial information that can be acquired through reading newspapers / magazine, watching TV and movies, searching the internet and via social media.

The above four groups A, B, C and D all add to the knowledge that the person uses as he executes the tasks.

E. INNOVATIONS

xiv) Innovations a key factor
As discussed in subsection 5.7, innovations boost knowledge, they alleviate levels in technology and knowhow that are in the knowledge that the practicing person uses.

If we try to see how we can quantify the effects of all the groups of factors A, B, C and D as well as E, the practice that is now in-use is what human resources departments use, that is following grades with categories for the pay scale of the employees. For a person's grade in a company, it starts with which grade he/she enters the company at, in accordance with his/her education, previous experience and in accordance with the requirements of the job he/she is to assume. Then the person's performance, contributions to the company is used to assess his progress; that translates into his salary increment(s), and bonuses and advancement(s) in the organization.

As was noted earlier, innovation leads to a higher level in knowledge in a similar manner that higher temperature leads to better performance (higher efficiency) in power plants.

Let us address how an enterprise assesses the internally achieved innovation. It is possible to track the costs that went into the specific R&D program from which the innovation came about. There could be rewards and incentives to those that made it possible. If the innovation resulted in producing a software or procedure that some staff members need to be trained on, the incremental costs for the specific training would be estimated.

Thus, even though the procedure is not straight forward, it is doable.

F. EXCEPTIONAL ABILITIES

xv) We see that there are among performers in sports, arts, whether individual or in groups, persons who reach exceptional levels, and they command very high compensations. Some professional athletes get multi-million USD compensation packages.

Also, some executive management positions command multi-million USD compensation packages.

From the considerations noted above under i) to xv), we can see that the work force compensation and pay scales as have evolved to date provide a reasonable starting point for coming up with the equivalent joules proposed in subsections 9.2 and 9.3.

KNOWLEDGE & INNOVATION COMBINATION (KIC)

One possible measure for the knowledge & innovation combination is a scale that relates to the time duration to complete a task or a set of tasks. Two time durations can be compared. TD1, before the introduction of a change in the knowledge with a certain innovation. TD2, after the change in knowledge with the selected innovation completed. The ratio of TD1/TD2 is a measure of the contribution of the knowledge change & accompanied innovation.The two cases 1 and 2 should have the same output or equivalent to have a meaningful comparison.The high TD1/TD2 ratio reflects a revolutionary innovation and the lower values echo an evolutionary innovation.

This knowledge & innovation scale may offer additional insights in the search for a dependable way to quantify the knowledge in Joules. The knowledge innovation combination (KIC) for 2, which generally, replaced the KIC in 1, has different characteristics with the K2 (knowledge at 2) at higher ranking than that of K1 (knowledge at 1). Thus, we need to factor the level of K2 to the level of K1.

The author went back to his earlier studies of thermodynamics and concluded that the thermodynamic term / concept "Exergy" renders a good means to handle this issue. However, the author had at the outset when he started this book adopted a rule "avoid complex thermodynamics in this book". So it suffices to say that the analogous term Exergy appears to hold a venue in the search of means to quantify knowledge in Joules. The author will deal with this subject in a future publication.

9.5 Interdisciplinary

From sections 5, 6 and 9 the interplays of knowledge and energy were underlined and demonstrated in several situations. For enterprises, we talked about electricity as the form of energy that is present in the K&E model application. In other words, in all the actions/tasks we can obtain through appropriate means, measurements or calculations, good estimates for energy.

For knowledge contributions, we can use a methodology that accounts for the factors numerated in subsection 9.4. Further, an additional consideration that is crucial in knowledge and innovation is the interdisciplinary effects.

Based on the author's experiences over the years, close interdisciplinary coordination is very beneficial. For instance, in engineering software development, it has been proven that an integrated team of the discipline specialist and IT professional working closely together renders timely development and gets quality results in the shortest time.

Looking into how other disciplines/fields do things can be enlightening and may lead to breakthroughs. This book is an example of combining knowledge from several fields.

9.6 Other Factors from Cost Perspective

Referring to section 6, when the models of the enterprises were discussed, there were two additional aspects beside the main consideration of the production phase:

I) Assets

The 1st phase of completion of the assets for the enterprise from commencement to completing the facility with all equipment that is needed for the production. This is applicable to both manufacturing as well as services. During this phase the enterprise is funding the capital needed to enable the utilization during the production phase. Please refer to Figure 6.1.1.

Relevant information on the investments of assets (to perform the pertinent computations), have been included in section 7 relative to economic factors such as the interest rate and section 8 devoted methodologies for investment evaluations from which the costs of the assets can be obtained.

II) Running Costs of the Facility

Please refer to subsection 6.1 (Figure 6.1.3). The related data can be obtained from historical data and in the case of a new facility, the data can be initially estimated based on the details of the facility and some data can be refined in subsequent years as appropriate.

9.7 Profit and Taxes

In section 9, the focus has been on the production phase of the operation of an enterprise, and in particular the use of the quantification K&E model. Additional costs were highlighted in subsection 9.4. To complement the discussions in the preceding subsections, to reach selling prices and to assess profits, it is necessary to consider depreciation and taxes. The marketing and sales department(s) under the guidelines set by the company management allow for profit margins.

Depreciation and taxes follow government applicable laws and rules and are estimated by the accounting department. The results produced by the accounting department are included in periodical reports. Management reviews these reports and orders actions as deemed necessary.

Further appropriate internal and external auditing is performed. The board of directors makes the final recommendation on the distribution of the profits for approval by the general assembly as applicable. These may have to be factored in the guidelines for the following year(s).

9.8 References & Bibliography

1) U.S. Bureau of Economic Analysis - https://www.bea.gov/data
2) U.S. Bureau of Labor Statistics - https://www.bls.gov

Section Ten:

BUSINESS ECONOMICS KNOWLEDGE & ENERGY

Over the last few decades, one can see a continuing trend towards ever increasing emphasis on quality, reliability, accuracy, and precision across the board in practically all the fields. Also, an emphasis on efficiency /performance, for instance in electrical power generation as was underlined in section 2. Further, the concern about the environment and in particular, global warming due to the CO2 emissions has emerged as a prime factor that is justifiably receiving considerable attention. For mankind, across the globe preservation of natural resources has become a major issue with profound effect on the welfare of future generations. In the author's view, the K&E interplay model should take a front and center position for all governments, private, old and young as all seek a better future.

For the author journey behind this book, really started in early 1988, when he observed during the economics course he took under his MBA courses, similarities of some figures with those he knew from thermodynamics – his field for undergraduate and graduate work is mechanical engineering. He collaborated with Professor Dr. Ibrahim Oweiss of Georgetown University on working out several ideas building on this view. The collaboration did not proceed at a pace both of us wanted primarily due to the excessive travels of the author on his job. In 2001, the two authors managed to complete the Reference. The author wants to express his gratitude to Professor Oweiss for his very valuable contributions to the Reference. The reaction from the economic community to the Reference in the early 2000's was not encouraging. It is only one Nobel Prize Winner, Economics, that provided a very valuable comment. Unfortunately, he passed away. The author left things aside and concentrated on his job overseas for 19 years. Of course, in the meantime he kept the subject of the Reference on his mind. When he retired at the beginning of October 2020, he saw that the Reference was visited about 8000 times on Google. That gave him an encouragement and his decision to write this book.

He opted to use a practical approach that brings the concepts to the reader through information readily available in various fields. Based on this approach, the 10 sections in the four parts of this book were compiled.

Parallel to the author's own journey, a lot led to many changes and as we concluded the second decade of the 21st century, we note real technology advances / digitization and environmental concerns e.g., CO2. We also see continued increase in percentage services versus manufacturing in the economy pointing to the contributions of knowledge. We saw in the last five years the emergence of the technology companies as the leading enterprises of the world as they reached unprecedented asset valuations.

10.1 Conclusions

It is very clear that the interplay of knowledge and energy is behind every activity in an enterprise. The same applies to government activities. Contributions of knowledge have taken front position in most activities in today's enterprises.

Section 3 shows the origin of knowledge, the human brains. While section 2 underlined the origin of energy is nature whether fossil or dispersed solar irradiation or blowing wind. In section 2 thermodynamic cycles / power plants were presented. Through these cycles low grade energy (chemical energy in the fuel to heat) is transformed to premium energy to form "electricity". Electricity can be put in use for a variety of purposes and can be transmitted through the TSO and the DSO networks to enterprises and residential users. From section 3, the human body systems transform low-grade energy in the food into premium energy that enable the person to do all his /her daily activities including his duties on the job. When we want to describe the capabilities that the person uses to perform his / her job, we find a big part of capabilities is knowledge that the person accumulates in his / her brain. The person draws on the knowledge stored in the brain to deliver on the tasks he / she performs.

It is noted that in the human body the knowledge and the energy pair's interactions are intimately taking place throughout the body with the blood carrying nutrition's and oxygen together with neurons in the brain & nerves system. The energy the different organs utilize is generated through chemical reactions as ATP's change into ADP's. These reactions take place on the molecular scale at the cells. This interplay in the human body is quite complex particularly when the storage in the memory in the brain is added. In the case of a power plant even with the computerized control system the knowledge and energy interplay can be seen in distinct parts of the plant. The important message is that the knowledge & energy pair is present and has pronounced effect in both cases.

Section 4 highlighted key technologies that have evolved in the last decade and now are at the forefront of the digitization movement. These are products of knowledge and innovations.

Subsections 5.5 Knowledge including data, information and technology underline the significance of the encompassing meaning of the word "knowledge". Added to this term is the term "Innovations" of subsection 5.6. This set is a key to all activities and the optimization processes of the pair knowledge & energy. Minimizing the input sum of knowledge and energy to reach the desired output is the goal of the optimizations every enterprise undertakes. This is behind the knowledge

& energy interplay model introduced in subsection 5.1 (as qualitative). The qualitative K&E model was applied for several enterprises in section 6.

After, the survey of macroeconomics (conventional economics) and investments in sections 7 and 8, respectively, the link between economics and the quantitative K&E model was established in section 9. The quantitative K&E model presented in subsection 9.3 is based on several findings and notes from the various sections 1-8 as highlighted in subsection 9.1. From subsection 9.2, "**knowledge can be represented by the equivalent Joules**". This is a key proposition. Thus, the sum of the joules of both the knowledge and energy can be added together. The sum can be minimized in search of the optimum solution for a product or a service. It is noted the energy uses in enterprises is mostly premium energy; that is "electricity".

The author believes that the term "labor" is now outmoded. It has served its purpose in the past, particularly when manual, and even in the eras when manufacturing was dominating economies.

For use of the quantitative K&E model the author believes that through historical data within the enterprise (specifics) the knowledge and energy data can be collected quickly. The procedures that have been applied in the examples of section 9 can followed.

The representation of the knowledge in the energy units as proposed in section 9 could open several developments in work force compensations, money, foreign currency exchange rates, inflation.

10.2 Revisit to Reference

Contrary to the approach that was taken in the **Reference**, which can be called a top –down approach, this book went the other way--a bottom up approach, starting from where we are, examining in some depth what we understand and building on that. This is the fundamental shift between the two approaches.

This book did not go into how low-grade knowledge can be raised to premium knowledge. Some of the issues in that direction were tackled in the **Reference.** The proposed representation of knowledge by joules in subsection 9.2 may bring us closer to what was sought in the **Reference.** In the Reference simulations from thermodynamics were borrowed to mimic economic situations. The author recognizes the difficulties of this approach for a non-engineer.

Nevertheless, some extracts from the **Reference** would be beneficial to the reader in grasping a key point; "enterprises are always in search and after higher level knowledge (technology)". This translates to a commonly known point: "higher level technologies bring higher values". This is now clear in the valuation of technology stocks. This was discussed in subsection 9.4: "Innovation and Knowledge Application" relative to the grades of knowledge. The approach in the **Reference** is a theoretical approach based on the analogy of terms in economics to their counterparts in thermodynamics, and the concept of "CAPITAL ENGINE" analogous to the "HEAT ENGINE" discussed in section 2.

Table 10.2.1 Economics Terminology versus Thermodynamic Terms

#	**Economics Term[E,1,3]**	**Corresponding Thermodynamics Term[2]**
1	Capabilities[E] [A] represent internal or latent attributes that a human being possesses, which enables that person to undertake an activity *or complete a part of an activity*[2]. It is a generalized of what we sometimes refer to as "Knowledge"	Energy "E"
2	Services[E] come about as a result of changes of Capabilities.	Changes in Energy
3	Utility[E] [U] is the total satisfaction derived from the consumption of goods or Capabilities changes as could be manifested through services.	Work "W"
4	Capital[E] consists of the durable produced goods that are in turn used in production. The major components of capital are Equipment, Structures and Inventory. In accounting and finance, "capital" means the total amount of money subscribed by the shareholders in the corporation.	Heat "Q"
5	Convertibility[E] Capabilities to Services; money is the term that resembles the convertibility.	Joule is the unit for conversion between heat and work or vise versa
6	Income Potential Function for Income[E] (IPFfI), while income is the flow of wages, interest payments, dividends, and other receipts to an individual or a nation during a period of time usually (one year), the **POTENTIAL FUNCTION FOR INCOME[E]** [I] that was proposed by the authors **Reference**, is a measure of usefulness. The higher the level of technology the higher the **IPFfI**.	Temperature "T"
7	Marginal Utility[E] [M_u]the additional utility arising from consumption of an additional unit of the commodity.	Pressure "P"
8	Quantity [N] is the repeat of commodity. The specific quantity is number of units per person.	Volume "V" Specific Volume "v" (volume per unit mass)

9	Labor is generally defined by the group of people of age of 16 and older who are either employed or unemployed. However, a more representative term that was introduced by the authors in the Reference, is **LABOR EFFORT**[E] [L] which depicts the effort exerted by humans. Specific Labor effort is Labor effort per person.	Entropy "S" Specific entropy "s" (entropy per unit mass)

Economic Term, superscript "E" designates an Economic term.

1) Thermodynamics Term. Please refer to section 2 for definitions.

2) BOLD TERMS WERE NEWLY PROPOSED BY THE AUTHORS **Reference**.

The two new terms may be difficult for an economist, but they were reached after extensive debates between the author and Professor Dr. Oweiss over almost ten years.

To supplement the above definitions, the following key equations are noted

Economics: $dSV^1 = I\ dL$ analogous to Thermodynamics $dQ = T\ dS$ Eq. 10.1

Economics $dU = M_u\ dN$ analogous to Thermodynamics $dW = P\ dV$ Eq. 10.2

In the **Reference**, a **CAPITAL ENGINE**[E] was introduced analogous to the heat engine (section 2). In the reference borrowing on known well established facts in thermodynamics and devising charts for I-L similar to T-S, and M_u – N to P-V. For the idealized Capital Engine equivalent to the Carnot cycle (Figure 2.3.1.3) found in section 2 the facts that we got from the second law of thermodynamics led to equivalent conclusions:

>>>> The higher I the higher the efficiency in the convertibility of capital – that is where technology under Knowledge come in.

1 Please refer to item 2 change appearing on the following page.

The above discussion forms a theoretical back-up to the discussions in subsections 9.2 and 9.3.

Further, referring to the suggestion made by the author in section 9, i.e., dealing with knowledge in the same units of energy, the following modifications to Table 10.2.1 are warranted:

In line# 2 in Table 10.2.1, the services or change in capabilities (knowledge) on the economics side, and on the thermodynamics side changes in energy will be "heat" and both services and heat will be quantified in Joules.

A correction for the line # 4 in Table 10.2.1 is due, the term "heat" on the thermodynamic side was not correct. In both economics and thermodynamics, it is necessary to deploy a setup that will be used in the production (Fixed assets - hardware and software etc.). E.g., in a factory, equipment etc., and in thermodynamics a power plant is built. In both cases Capital is expended.

In line # 5 of Table 10.2.1, once joules are used for both knowledge and energy, implications will follow for money. The author notes that in section 9, he used money to complete the demonstrative examples.

It is the opinion of the author that the contents of this book, and specifically from subsections 9.2 through 9.4 with the use of historical data as suggested in these sections, will open the way to attain the proper representation of **POTENTIAL FUNCTION FOR INCOME**^{E} [I] and Labor Effort [L] and Capital [K]. Once this moves forward significant advances in economic theories could come about. Finally, economists could be convinced with the merits of the material in the **Reference**. This was behind the second line in the title of the Reference "New Horizons, Shifting the Paradigm".

The author would like to note the contents of part 4 of the **Reference** "Human Model – What Next" as it contains useful insights for the term Human capabilities and Knowledge and the essence of the knowledge exchanges.

The author will not go further on the material in the **Reference**. Interested readers can refer to the **Reference**.

The Reference did not address the thermodynamics concept "Exergy". Simply this concept is hard to follow even among engineers. In the Reference, the basic economic and thermodynamics analogies were covered as noted in the Table. 10.2.1. Very briefly, the concept of Exergy uses temperature and changes in entropy. Exergy is measured in Joules. When the author started this book, he adopted a rule avoid discussions of complex thermodynamics. He is sticking to this self-imposed rule.

The author will address the analogy of Knowledge & Innovation Combination KIC to Exergy (subsection 9.4) in a separate publication.

10.3 Multi-disciplines

The author notes it is his conviction that other fields different from the FIELD one is familiar with and has practiced in could potentially provide very useful insights that are not directly available in the FIELD. Through looking outside the known bounds of the field / or collaborating with others from a different field, significant beneficial and useable information are obtained.

As an example, few years ago, after completing my dissertation, I started getting familiar with one of the early lasers that were produced by the company Spectra Physics. I was discussing my new experiences with an eye surgeon, asked me if there could be a way to measure the temperature at the inner surface of the eye if he would apply a laser to shoot at the defective spot without burning other neighboring points. Such dialogues stays with a person like me, and has always made me keep my eyes open and observant.

Throughout my career, I learned and applied things from other disciplines. My work in project development since the late 1990's exemplified the importance of multi-discipline collaboration.

10.4 Overall Perspective for an Enterprise

A fundamental message is clear. Trade is based primarily on knowledge. Innovations are key to raising the level of knowledge that is reflected in the items to be sold. The enterprise needs to invest in fixed assets that will be utilized in the preparations of the deliverables that will be sold. For the preparation of the deliverables, the enterprise, whether a manufacturing or services or mixed characteristic, all the tasks / actions (through outsourcing) involve knowledge and energy. The enterprise strives to minimize the cost of the combinations of the knowledge and energy. This overall concept should always guide the management of the enterprise. Please refer to Figure 1.3.1.

For any enterprise, there always consideration of the following:

- v Minor Investments – for sustaining continuous improvements
- v Moderate Investments – for step modifications in transforming the deliverables(s)
- v Major Investments – for substantial changes of the deliverables or introducing additional deliverables.

In all the above three types, considerations of knowledge & energy in both the fixed assets set up as well as the production phase are musts. In the background are the information such as those in sections 7 and 8, Macroscopic Economics and Investments, respectively.

10.5 Recommendations

1) The author hopes that the economic community be open minded in seeing a new / different approach.
2) It is suggested that researchers investigate further the knowledge quantification; the author believes that the equivalent Joules is a good step. The author recommends investigating the quantification of knowledge by joules in the various technologies of information services such as those cited in section 4.
3) Researchers need to examine the grades of knowledge and the effect of innovation.
4) The author feels that the term labor is now obsolete and amid the predominant role of knowledge, it needs to be replaced by terms that are more representative of what most workforce occupants do in their work.
5) Researchers, economists, and engineers are encouraged to explore combining the concepts of the **Reference** and the results of the practical approach taken in this book towards reaching further understandings of how economy and thermodynamics have much in common.
6) The author learned a lot as he was researching for information to prepare section 3. There is already very valuable information that was not there two decades ago. Medical researchers can add a lot through further research on the brain and its operations with respect to storage of information and control of the human organs. More and deeper investigations of the analogies discussed in this book between the human body / brain and the electrical system / controls in sections 2-4 are recommended.
7) This book brings to light many insights into knowledge as the most significant factor in the economies and why services now dominate economies. Additional work pertinent to recommendations 2-6 is strongly recommended.
8) For the leading technology companies sponsoring research in various areas related to topics covered in this book and the **Reference** will accelerate advances and development of additional new information.

ABOUT THE AUTHOR

Dr. Hemmat H. Safwat Dr. Hemmat H. Safwat is an Energy Expert. After a professional career that spanned over 50 years in technical and managerial capacities, he retired in September 2020 and currently serves as a consultant for power and desalination plants and renewable energy.

Dr. Safwat was in academia between 1964 to 1973, where he taught and carried out research at Cairo University, Egypt, West Virginia University, USA and Delft University of Technology, the Netherlands. During that time, he completed his M.Sc. in Solar Energy Utilization, Ph.D. Radiation Heat Transfer, D.Sc. in Fluid Mechanics.

In 1974, Dr. Safwat joined Bechtel, San Francisco and initially was a technical specialist and evolved to supervisor and to a project manager position. He was at Bechtel for 23 years, participated in energy technology evaluations, mechanical analysis for nuclear power plants systems.

Between 1986-1987, Dr. Safwat was on a leave from Bechtel and served as a Visiting Professor in Mechanical Engineering at the AUC (American University in Cairo). In 1988, Dr. Safwat obtained a MBA from the Golden Gate University, San Francisco.

In 1989 he assumed a position in Cairo as Bechtel Project Manager for fossil power plants located in Egypt as well as planning for the nuclear power plant at Dabaa. He made significant contributions to the Egyptian Ministry of Electricity. Upon returning to the USA, he was Bechtel's Project Manager for a nuclear power plants' improvement program in central USA.

In June 1997, Dr. Safwat joined Enron and served as a General Manager for the development of power and desalination plants for Enron International – Middle East. For five years he managed the development of independent power plants IPPs and Independent power and water plants (IPWPs) in several countries in the region.

In late 2001 Dr. Safwat joined Consolidated Contractors Group offshore "CCC" (at the managing office in Athens, Greece). Dr. Safwat led the power plants bidding both for the EPC (Engineering Procurement and Construction) and development projects. The projects included power plants in the Middle East and Africa. Dr. Safwat also led in 2011 the formation of CSPC (Consolidated Sicon Power Company), a fully owned company by CCC that handles EPC power including renewable energy PV (Photo Voltaic) and wind.

Dr. Safwat is an ASME (American Society of Mechanical Engineers) Fellow.

Dr. Safwat participated in organization of numerous professional conferences. Dr. Safwat has more than 50 publications. He served on ASME committees and chaired sessions in numerous conferences. In 2002, he co-authored with Professor Ibrahim Oweiss the book *Economics: New Horizons, Shifting the Paradigm*.

WA

Made in United States
Cleveland, OH
06 November 2026

47771491R00129